HOW FLOWERS WORK

*Frontispiece: Cross-leaved heath (*Erica tetralix*)*

The publishers regret that the caption on this page has been included in error, and that the photograph on page 83 has been inverted.

BOB GIBBONS

HOW FLOWERS WORK

A GUIDE TO PLANT BIOLOGY

Drawings by
Vanessa Luff

BLANDFORD PRESS
POOLE · DORSET

First published in the U.K. 1984 by Blandford Press,
Link House, West Street, Poole, Dorset, BH15 1LL.

Distributed in the United States by
Sterling Publishing Co., Inc.,
2 Park Avenue, New York, N.Y. 10016

British Library Cataloguing in Publication Data

Gibbons, Bob
How flowers work.
1. Plant physiology
I. Title
581.1 QK711.2

ISBN 0 7137 1278 3

Typeset in 10/13 Plantin light by Asco Trade Typesetting Ltd., Hong Kong.
Printed in Hong Kong by South China Printing Co.

CONTENTS

The familiar snowdrop flowers very early in the season, sometimes well before winter is over, and long before the trees are in leaf.

INTRODUCTION

Plants are the whole basis of life on earth. Without them we could not survive, yet we take them completely for granted. If we cut the lawn, it regrows; if we plant bean seeds, they turn into plants covered with more beans, and every year in late winter the snowdrops push up through the cold soil to produce their delicate white flowers. At the same time the mantle of every landscape we admire is made up of a mosaic of plants, which may include the golden colours of autumn birches, or meadows bright with flowers, or the dark green of a pine wood.

Most people simply accept these things without giving a second thought to the questions they raise: why do snowdrops flower in late winter and not in summer, and how do they know it is late winter? Why do other plants only flower in autumn? What makes seedlings or flowers grow towards the light, and why do roots grow down through the soil instead of up into the air? Why does a hedge go bushy if we cut it? Perhaps the most difficult of all, why do the few tiny cells of, for example, a lettuce seed always develop into a lettuce, rather than something else? Where is all the information that tells them what to do?

It has gradually been appreciated over the last 50 years or so that plants and vegetation are incredibly complicated things, and that they share a surprising number of characteristics with mammals. Plants, for example, have a complicated internal hormone system which controls many of their internal functions, and, as in animals, involves a number of compounds which may work together or against each other. It is now known that some plants can measure time very accurately, to within a few minutes in some cases, and that they time their flowering according to the length of day. It has even been shown recently that some plants have a memory, retaining information about previous events that happened to them.

Not only are plants themselves fascinating and complex beings, but so are the communities that they live in—the oak forests, the grasslands, and so on. None of them are as stable as they look, but are made up of hundreds of species, all competing together in slightly different situations for available light and nutrients, and changing all the time. A grassland may appear to be an unchanging community, but we now know that there are little detail changes going on all the time as plants die off and are replaced by others of the same species, or something quite different, and gradually the whole character of the grassland may change.

This book attempts to explain some of these mysteries and answer some of the questions, though it has to be admitted that plants have by no means yielded up all their secrets yet!

1 · THE EVOLUTION AND CLASSIFICATION OF PLANTS

It is now generally believed that there was no life at all on earth until almost 4,000 million years ago. So, unless one believes in instant creation or extra-terrestrial interference, we have to accept that the bewildering array of life forms on earth today, from bacteria to forest trees, ants to elephants, all evolved from a purely chemical origin. Over the 3,500 million years or so that life has been present on earth, there has been a continual changing of forms as new types have arisen and old ones improved, and this process is known as evolution.

The first step seems to have been for water to combine with methane and ammonia, under the dynamic influence of sunlight and perhaps lightning, to form amino acids, the building blocks of proteins. As there was no atmosphere to speak of then, the ultra-violet rays were considerably stronger than now, and their effect on chemical reactions was probably dramatic. Although it may seem unlikely that such a step, from nothing to life, took place, something similar has been recreated in a laboratory. At some subsequent stage, the amino acids formed into proteins, and then an all-important membrane must have appeared to allow the first cell to be formed. The very first organisms were probably like viruses, only free-living; that is to say combinations of chemicals with little semblance of life except the power to reproduce themselves.

The first cellular organisms were probably rather like bacteria, making a living from the 'primeval soup' by a process similar to fermentation. There are records of fossilised plants, resembling blue-green algae, from rocks dating from 3,100 million years ago! This find is highly significant for two reasons: firstly, blue-green algae still exists today; secondly, the cells resembling them had almost certainly started to photosynthesise, and to release oxygen as a byproduct into the atmosphere. Without these plants there would have been no oxygen at all, and not only did this ultimately allow other life forms to develop but it also protected the developing life forms from excessive ultra-violet by forming in the upper atmosphere as ozone.

So, although the atmosphere of the earth some 3,500 million years ago would be poisonous to almost all living things today, its components were necessary for life to form, and it would be impossible to recreate life in the same way from the present atmosphere.

Blue-green algae are very primitive and very unusual plants—their cells have no nucleus, for instance—and it was not until about 1,500 million years ago that something more akin to the majority of present day plants arose. These were the green algae, like the green 'slime' that grows in aquaria when they are in the light. These cells had nuclei, and they gave rise to the possibility of multicellular organisms with several cells working together. Nevertheless, for a while after this there is little visible in the fossil record—as tiny organisms leave little behind when they die—but we know that by the beginning of the Cambrian period, some 570 million years ago, there was an enormous range of life forms present on the earth. Over 900 different recognisable species have been identified, and when it is realised that only a tiny fraction of living things become fossilised, let alone preserved for 500 million years and then found again, it can be seen that there were probably many more species around at the time. The species identified from this period include, among other things, types that are clearly similar to present-day fungi and seaweeds.

Up until about 400 million years ago, such organisms as there were lived entirely in a watery environment. In the Silurian period (435 to 395 million years ago) another major advance occurred, as plants 'emerged' from the sea and began to colonise the land. The early land plants included forms like present-day mosses and liverworts, as well as numerous types long since extinct. To achieve the change to the terrestrial environment, they gradually developed roots to absorb the water that had been so abundant, and their methods of reproduction became less dependent on water. It is interesting that many of the present-day groups that relate most closely to these early land invaders still have a need for water and a humid environment for reproduction to take place, whereas the more advanced flowering plants have mainly lost this requirement.

By the time of the Coal Measures, the Carboniferous period (about 300 million years ago), there was an enormous range of land plants, including the giant horsetails, tree ferns and clubmosses, and it is their remains that form the coal we use today. Many of these forms were lost in the succeeding Permian period, which was a particularly dry time throughout most of the world, but those that survived continued to evolve and eventually gave rise to the seed ferns (now extinct, but with some of the characteristics of today's seed plants). It was another big step forward to produce a seed with all the food reserves necessary for subsequent germination and survival, and from their appearance some 200 million years ago the flowering plants with their enclosed seeds have risen to a position of dominance in the vegetation of today.

THE PROCESS OF EVOLUTION

Why should all these changes have taken place, and why are there so many hundreds of thousands of different species of plants today? The answer lies in the process of evolution by natural selection, as first described in detail by Charles Darwin.

One cannot fail to notice that human beings vary widely, and that offspring are not identical to parents. Almost all organisms share this characteristic of producing variable offspring, and since most plants produce large numbers of potential offspring it follows that the progeny will have a wide range of characteristics. Each will face the fight to survive in a hostile environment, and only those that are best fitted will live long enough to reproduce; most of the others will die sooner or later before reproducing. Gradually the characteristics of the population will change, over the generations, moulded by the environment it is living in. For example, if one offspring of a plant has flowers which are particularly attractive to insects, its flowers will be pollinated first and it will produce the most seeds. Many of its progeny will probably share this characteristic and, if it continues to be successful, the proportion of such flowers in the population will continue to increase. Today we only see the results of successful evolution, for the majority of forms and potential new species have not survived.

Evolution does not go in any one 'direction', although it may appear to when seen in retrospect, but instead produces a multiplicity of possibilities, any one of which could fit the needs of the moment. The net result, over millions of years, has been to produce the enormous variety of plants found today, the survivors best fitted to the present environment. Evolution still continues and new species are gradually arising, though many others are unable to evolve rapidly enough in the face of man-induced environmental changes and so they become rare or even extinct.

One intriguing aspect of evolution is the way in which widely separated, unrelated, groups of plants may evolve the same strategy to cope with a particular problem, and may come to resemble each other very closely. This is known as convergent evolution. For example, many unrelated plants growing in deserts throughout the world have evolved in a cactus-like form, suggesting that it may be the most suitable solution to the problems they face.

THE CLASSIFICATION AND NAMING OF PLANTS

Because there are so many different types of plants throughout the world, or even in one place, it is essential for scientific, economic and common sense reasons that they are classified and named. Rather than simply naming plants and grouping them according to, say, flower colour, the whole basis of plant classification rests on botanists' perception of how plants are related and how they evolved to their present state, and plants that are grouped together are believed to be very closely related, with a common ancestor.

Until the eighteenth century, plants were 'named' simply by Latin descriptions of them which could run to several hundred words. The great Swedish botanist, Carl von Linné (or Linnaeus) revolutionised the system and provided the basis for the modern classification of all animals and plants by proposing that each organism should simply have two names. The first name is the generic name, which a number of closely-related plants share, and the second is the specific name. Within a genus, no two plants have the same specific name. For example, the generic name of the buttercups is *Ranunculus*; to distinguish the various different species within the group they are all given different specific names such as *Ranunculus acris*, the meadow buttercup, *Ranunculus bulbosus*, the bulbous buttercup, and so on. Although the specific name may be used elsewhere in the plant kingdom to describe other plants, e.g. *Poa bulbosus*, one of the meadow-grasses, the combination of the two names is unique to one plant.

Linnaeus provided many of the names still in use today, but as our knowledge improves and ideas change, so names may be altered, either so that a plant is put into a category to which it is better suited (for example, if detailed studies had revealed that one of the *Ranunculus* species was really more closely related to the anemones), or when it is discovered that a botanist has previously given a plant a different name, which may then have to be adopted. These double names are known as binomial names, and each one refers to the basic unit of classification—the species.

A species is a group of populations of plants that have definite and constant characteristics in common, and which usually breed amongst each other to produce fertile similar offspring. We have seen already that species vary within themselves (the basis of natural selection and evolution), but they vary within limits and do not normally overlap with closely-related species. The definition is a useful working one, and most people know what is meant by it, but it does not always hold true; sometimes, two species hybridise to produce fertile offspring, and other species grade imperceptibly into each other. These are the problem groups, often still evolving, but the majority of plants are easy enough to identify and classify as a particular species.

Similar species are grouped together into genera; for example all the different buttercups comprise the genus *Ranunculus*. Closely related genera are grouped together into families which may contain hundreds of species between them; for example the members of the dandelion family, the Compositae, generally share the characteristic of having a particular compound type of flower, yet there are over 14,000 species in the family worldwide. Families are grouped together into orders, though sometimes intermediate groupings such as tribes, subfamilies and so on are used. Orders are grouped into classes, and classes are grouped into divisions or phyla, which form the major groups of the plant world such as bacteria, fungi, or the ferns and horsetails.

THE PLANT KINGDOM

The whole kingdom of plants encompasses everything from forest trees to microscopic algae, together with a few others that are somewhat doubtful. It is impossible to describe the whole range of plants here, but a brief summary of the main groupings may be useful, starting with the most primitive (i.e. those that have evolved least from the primeval plant types), and ending with the most advanced—the flowering plants.

Bacteria

Bacteria are generally considered to be more closely allied to plants than to animals, though they really have little in common with either group. They are microscopic single-celled organisms that multiply by simply dividing into two identical cells, at a prodigious rate in favourable conditions. One bacterium cell can produce over 32 million bacteria in a single day, given ideal conditions. Many of them cause diseases in both plants and animals, yet many others are useful to us. Some live in mammalian digestive systems and help to digest substances such as cellulose; others produce yoghurt from milk, while others live in association with higher plants (such as broad

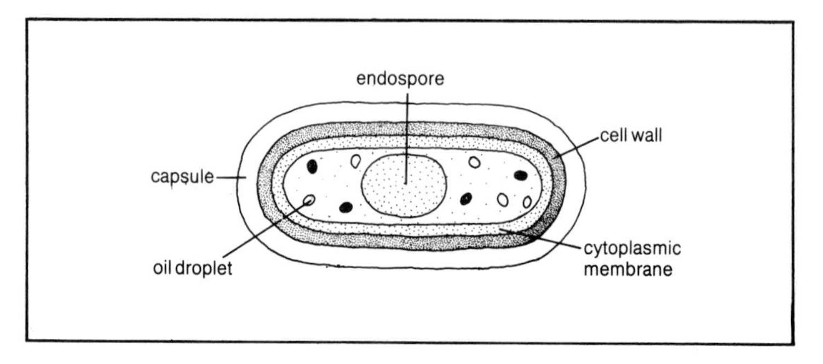

Diagrammatic representation of a capsulated bacterium, without flagellae (hairs).

beans) and produce nitrates from the air which the plant can then use (*see* Chapter 3).

None of them can produce their own food by photosynthesis (*see* Chapter 3), and most of them live as parasites or saprophytes on dead organic material. A few have the special ability to live on inorganic substances such as ammonia, sulphur or nitrogen, and (since the early atmosphere of the earth contained considerable quantities of ammonia) they are considered to be direct links with the earliest plants on earth.

Viruses are even smaller than bacteria and consist of just a few molecules of protein, but there is no particular reason to classify them as plants.

Blue-green algae

Blue-green algae are a peculiar group of organisms with little relationship to true algae. They are of particular interest since fossils found in rocks some 3,100 million years old closely resemble modern blue-greens, indicating that they are extremely primitive. Although they are able to photosynthesise, their cell organisation is quite different to algae and all higher plants in that they have no nucleus in the cell (*see* Chapter 2). They normally occur as single minute cells or small groups of cells, but some forms are clearly visible to the naked eye; for example, the yellowish-green slime that grows on some paths or lawns in winter consists of colonies of the blue-green alga *Nostoc* encased in slime.

Fungi

Fungi are simple plants without the specialised cellular organisation of the higher plants, though they have peculiar features of their own. Many fungi

*The cep or penny bun (*Boletus edulis*) shows the familiar form of a fungal fruiting body with the spore-producing cap.*

have a substance called chitin in their cell walls, as well as cellulose; chitin is a characteristic component of animals, especially insects, rather than plants. Unlike most plants, all fungi are unable to manufacture their foods by photosynthesis as they do not possess the green pigment, chlorophyll.

When most people think of fungi, they think of mushrooms, but mushrooms and similar fungi are simply the fruiting bodies of a small group of fungi. Other fungi include the mould that grows on stale bread, black spot on roses, and yeast. Most fungi have a characteristic structure of thread-like bodies called hyphae, which aggregated together form mycelia. Fungi such as mushrooms have extensive mycelia growing below the soil, or in rotting wood, and the mushrooms themselves are simply fruiting bodies thrown up at a particular time of year to produce and disperse spores.

Since they cannot normally produce their own food, fungi live either as parasites on other plants, or occasionally animals, or they live saprophytically by making use of rotting organic material such as dead wood,

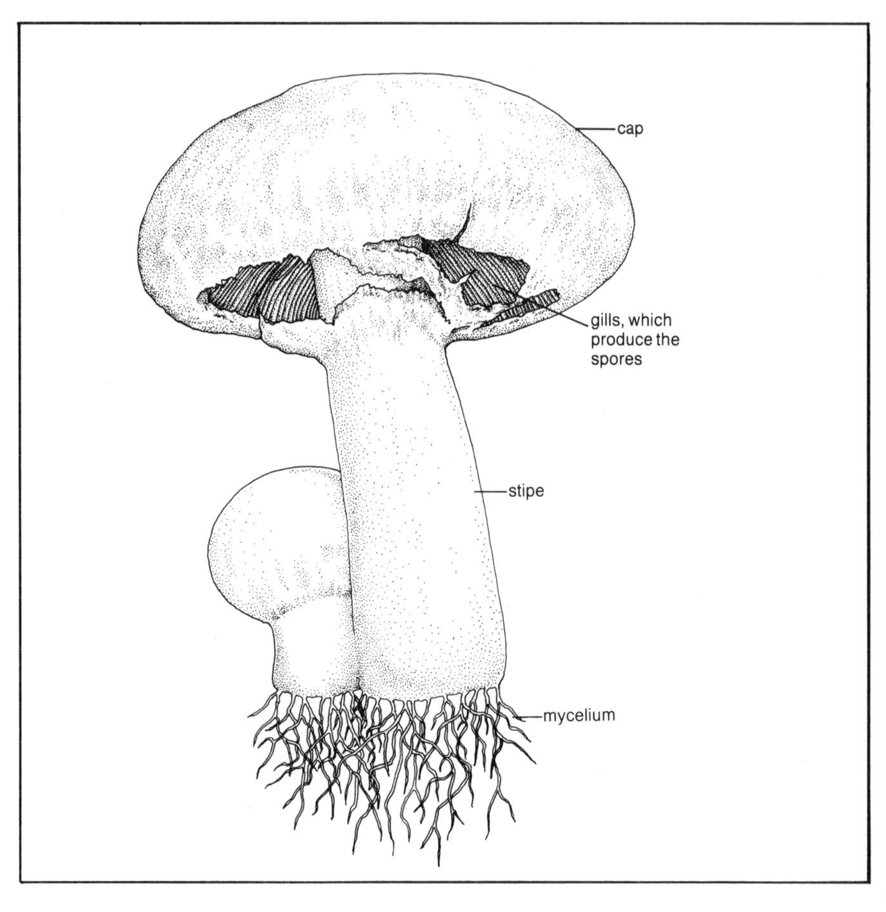

A characteristic fungus fruiting body with below-ground mycelium shown.

leafmould or stale food. Yeast is an exception in that it can convert sugar into alcohol and carbon dioxide by the process of fermentation, used throughout the world to produce alcoholic drinks. Other economically useful fungi include the blue mould *Penicillium notatum* from which the drug penicillin was first isolated.

One group of fungi, the slime-moulds, are capable of movement, sliding considerable distances over soil or lawns.

Algae

Like fungi, algae are primitive plants without any specialisation into leaves, stems, flowers etc. Almost all, however, share the characteristic of being able

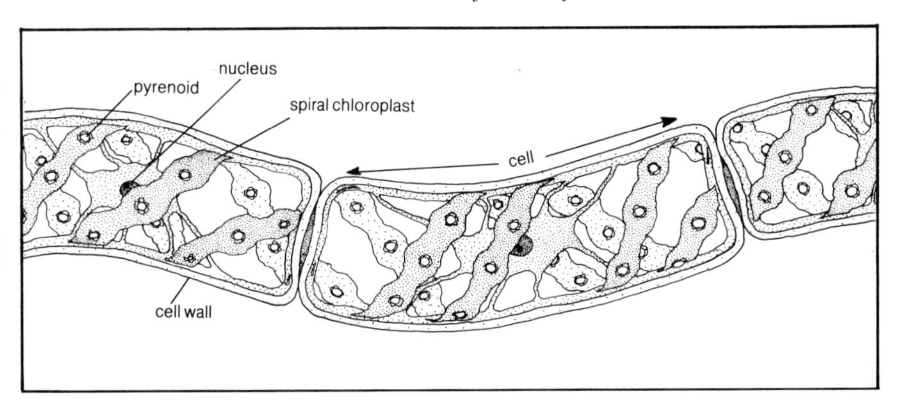

Enlarged cells of filamentous alga, e.g. Spirogyra.

to manufacture their own food, in the presence of sunlight, by photosynthesis. Although most people know little about algae, some forms are familiar to everyone, as the whole range of 'seaweeds', including brown algae, red algae and green algae, are all in this group. Hardly any higher plants have colonised the sea, and only very few fungi, so almost the whole of the marine vegetation is dominated by algae. They have developed different colours, which often mask the basic chlorophyll green, because the character of light changes as it passes down through sea water, and different pigments are required to absorb what light there is at depth. Near the surface, green algae thrive, absorbing more or less unaltered sunlight; lower down, where the light is more green and blue, brown and red algae do best.

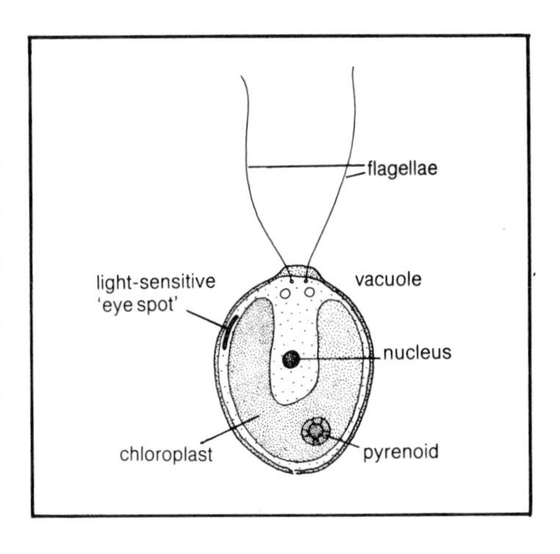

Single-celled mobile alga.

Despite their lack of specialised tissues, some marine algae can grow to enormous sizes because of the buoyancy of the water and the continual presence of adequate nutrients, water and light. The giant kelp of the Atlantic grows to lengths of up to almost 600 feet (183 metres), at depths of 100 feet (30 metres) or more below the surface. It is not normally seen except by divers, but bits occasionally break off and subsequently appear on beaches after storms.

Less well-known are the terrestrial algae. The green scum on ponds, or the green film that develops on aquaria glass in the light are both examples of green algae. Some are single-celled, whereas others are filamentous or colonial. A few, like the Euglenophytes, share some characteristics with animals: firstly, they can propel themselves by means of movable whips or hairs called flagellae; secondly, they have a red eye-spot which is sensitive to light and allows them to find the most suitable conditions to allow them to photosynthesise; and thirdly, they are able to ingest (i.e. eat) particles of solid food. They are generally considered to be plants, however, especially in view of their ability to photosynthesise.

The greenish-grey powder that often grows on tree bark, especially in damp places, is another unicellular green algae, known as *Pleurococcus*. Another group of algae that are occasionally noticed are the stoneworts or Charophytes which appear to have the structure of flowering plants with stems and branches. They are fairly common in calcareous ponds or slow-moving rivers which are clear enough to allow light penetration.

Lichens

Lichens are amongst the strangest of plants, yet most people are barely aware of what they look like. They exhibit a wide variety of forms, from orange or grey crusts on rocks or walls, to hanging beard-like plants growing on tree branches, or grey wine-glass shaped structures with red knobs on them! Despite the fact that there are about 16,000 species of lichen throughout the world, they are all made up of a unique union between two quite unrelated plants—an alga and a fungus. Different species of fungus have combined with different species of algae to form individual constant species of lichen, able to reproduce themselves as lichens. They are all composite plants, with two quite distinct species growing together in a union that is obviously favourable to both. Although it is not particularly surprising that two species should grow together for their mutual benefit, it is surprising that each union gives rise to such a persistent, self-replicating, highly-individual species. It is hardly surprising that for a long time the dual nature of lichens was not recognised at all.

Lichens, such as this Cladonia polydactyla, *are compound plants made up of an alga and a fungus which gives them the ability to colonise many inhospitable places such as bare rock or wood.*

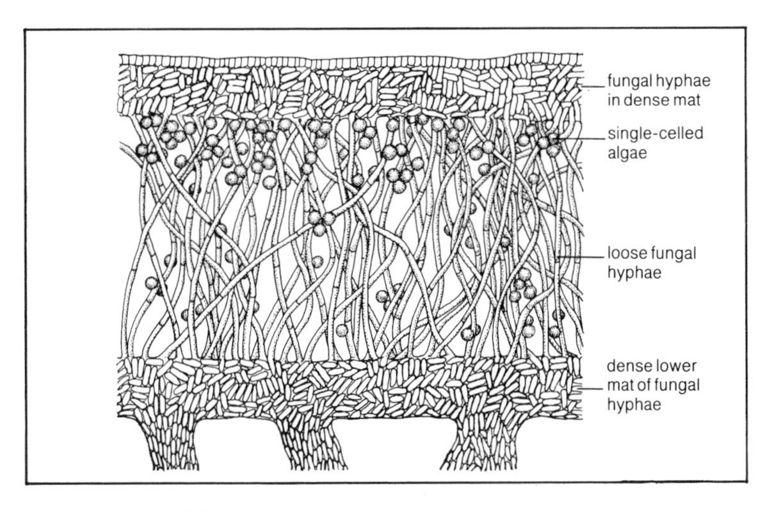

Cross-section of lichen.

The combination of two quite different plants, with differing methods of nutrition, has allowed most lichens to become highly efficient colonisers of bare ground, rocks, tree bark, and other areas where there is little competition. The algae can photosynthesise, and the fungi can spread through the substrate absorbing dissolved nutrients. At the same time, the lichens are able to secrete acids which can even dissolve rock, and allow them to get a foothold where there is no obvious source of nutrition. They are also very tolerant of cold, and in polar regions or near the snow line they may often be the dominant, or only, form of vegetation.

Because they grow slowly and regularly, they can sometimes be used to date other objects—the time a roof was built, the time a rock has been split from a cliff, or the length of time since a glacier retreated exposing the rock below, though the technique is not as simple as it sounds. They are also used as a source of dyes, as indicators of air pollution, as food (in Iceland), and the famous reindeer moss (the main food of reindeer) is actually a lichen, *Cladonia rangerinifera*. In Britain, at least, they are now being used to assess whether a woodland is ancient or not, as judged by the collection of lichen species that occurs in it.

Bryophytes: mosses and liverworts

As we have already seen, mosses and liverworts are primitive land plants, not dissimilar to the early colonisers of the land, though they have continued evolving ever since. They are more advanced than algae or fungi, and the mosses in particular have a definite division into stems and leaf-like structures; they also have the beginnings of a vascular conducting tissue, though this is not nearly as complicated as that in the flowering plants. Unlike flowers and conifers, they reproduce by spores rather than seeds, and this involves a cycle known as the alternation of generations. The plants which we see and know as mosses or liverworts are one generation, with a single set of chromosomes (a condition known as haploid). These produce male and female cells which eventually fuse and produce a cell with twice as many chromosomes (known as diploid). This grows into the other generation, which normally remains attached to its parents; this second generation produces a capsule containing spores, each of which now contains the original number of chromosomes, and these can then grow into new haploid plants when conditions permit. The whole process is very dependent on the presence of water and high humidity, a relic of their aquatic origins and one of the barriers to bryophytes becoming more widespread. Ferns exhibit a similar alternation of generations, except that the two generations are completely separate.

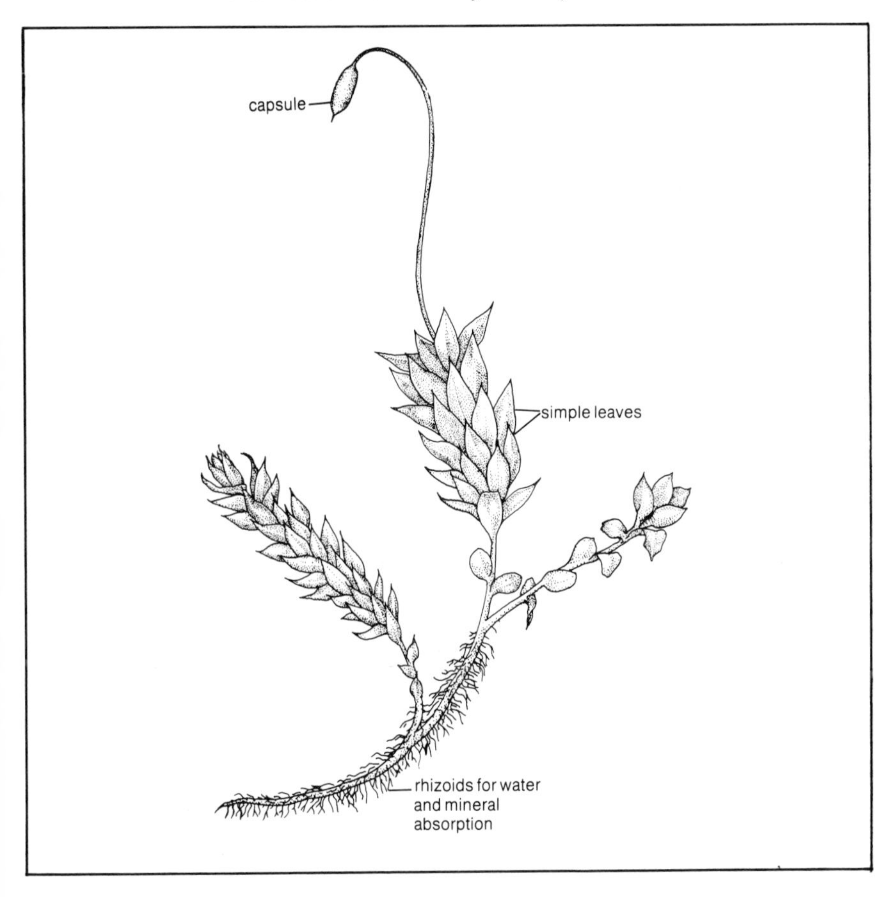

Structure of a typical moss (sporophyte generation).

Pteridophytes: ferns, horsetails and their allies

Ferns and their relatives are considered to be primitive in that they reproduce by spores rather than seeds, and they still exhibit the alternation of two generations with its dependence on water or humidity. In other respects, though, they are much more advanced than the mosses and liverworts with their complex vascular tissues, allowing a much larger plant to develop. Ferns are successful plants, common throughout much of the world, yet most of them are confined, by one weak link in their life-cycle, to humid or damp situations. In the mosses and liverworts, the dominant generation is the haploid sex-cell-producing generation, the gametophyte and the spore-bearing generation being dependent on this. In the ferns, the spore-bearing

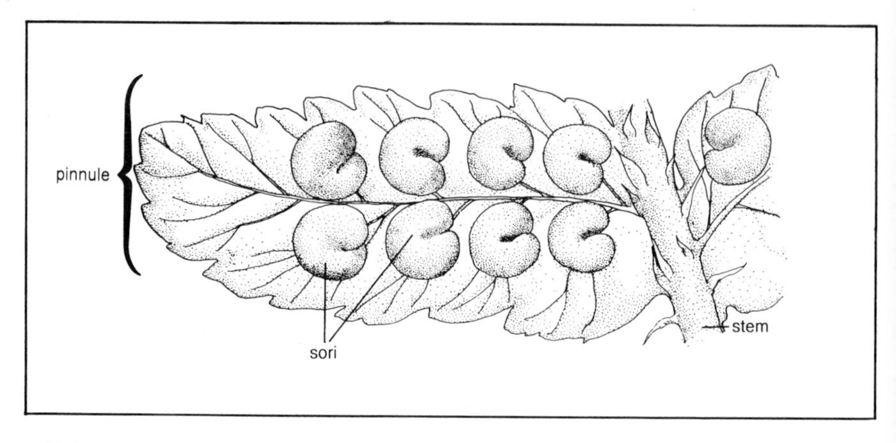

*Enlarged diagram of the underside of a small section of a fertile fern frond (*Dryopteris *sp.) Spores are produced under the heart-shaped covering of the sori.*

generation has become separate and is the dominant part of the life-cycle. The ferns that we see (the bracken, hart's tongue or maidenhair fronds) are only half of the cycle; they produce spores, normally in special organs on the backs of the leaves, and these spores disseminate, and grow into a completely different plant, rather like a liverwort, and known as the prothallus. This prothallus is haploid, i.e. it has a single complement of chromosomes, and it gives rise to the male and female cells. These fuse, and the resulting fertilised cell grows into a new, diploid sporophyte—the fern plant—while the prothallus dies.

In evolutionary terms this is a considerable advance, as the sporophyte is able to adapt to a wide variety of conditions; the weak point in the cycle, compared to flowering plants, is the prothallus stage which requires water and humidity, and this of course determines where the 'adult' plant grows. As we shall see, the seed plants have overcome this problem, and a few of the more advanced fern allies, especially the clubmosses, have evolved a rather similar system. The horsetails, too, are relatives of the ferns. For the time being, they have 'had their day' for they were the dominant plants at the time the Coal Measures were laid down, and indeed most coal was formed from the remains of horsetails and their relatives growing in extensive

*Unlike flowering plants, ferns such as this hart's tongue fern (*Phyllitis scolopendrium*) reproduce by spores produced in sori, visible here as lines under the frond.*

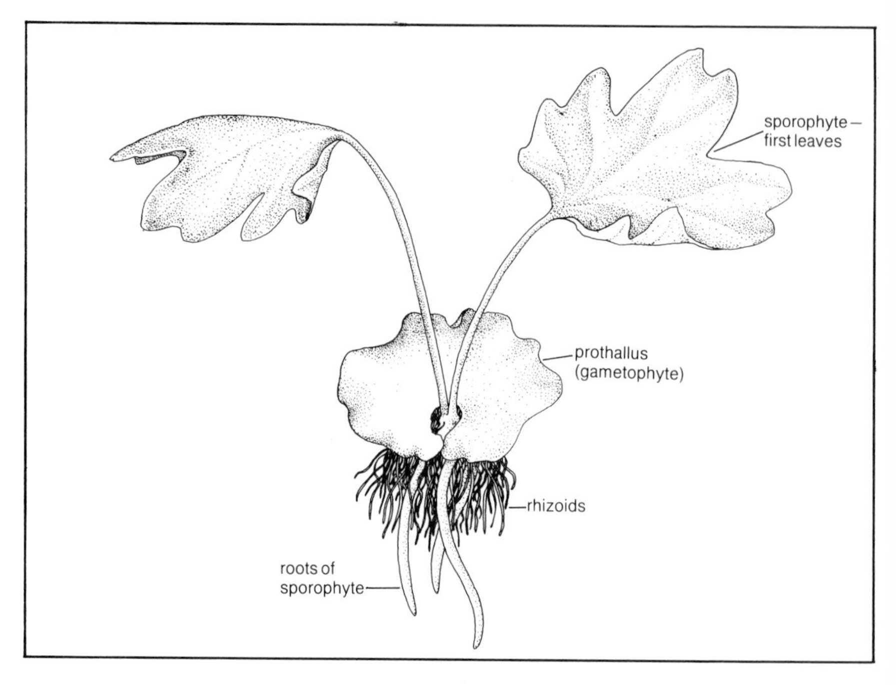

The young sporophyte of a fern developing from the gametophyte prothallus.

swamps. Today, they are widespread but rarely abundant or dominant, though one—the field horsetail—is a common and widespread weed of farmland and gardens.

Seed plants: conifers and flowering plants

All the plants that we have so far considered reproduce by means of spores. These are tiny objects often consisting of only a few cells, with no food reserves, produced directly on their parent without the fusion of sex cells to form each one. This method of reproduction is far from ideal, as the spore needs ideal conditions to grow in immediately it germinates, and a high degree of humidity is required. About 350 million years ago, however, the first plants bearing true seeds evolved. These were the seed ferns, and they lasted for about 50 million years when their place was taken by the gymnosperms. The gymnosperms (the conifers and their allies) bear their seeds without any protection, that is they have naked seeds. The flowering plants, however, which evolved later, have their seeds protected against the elements by a tough outer covering, and they represent the most evolved form of plants found on earth.

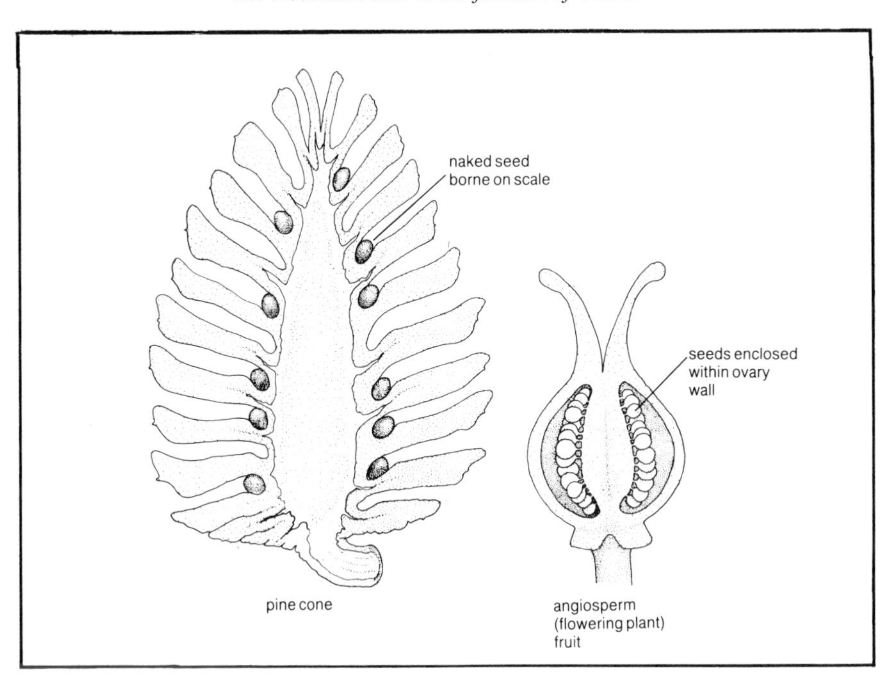

Cone of a gymnosperm (pine) compared with fruit of an angiosperm in section.

Both the gymnosperms and the flowering plants represent a major advance over their evolutionary forebears since they have avoided the difficulties of alternating generations and the requirement for water by, in effect, retaining the gametophyte generation within the parent plant, and producing seeds. The seeds contain enough food reserves within them to give them a start in life whether or not conditions are ideal at the time.

Amongst the flowering plants, one further major classification should be noted—the division into the monocotyledonous plants and the dicotyledonous plants. The cotyledon is the leaf within the seed, and 'dicots' have two such leaves whereas 'monocots' have only one. When most seeds germinate, whether they are herbaceous annuals, broad beans, or weeds, the resulting seedlings have two tiny identical leaves, which often differ markedly from the leaves the plant will develop later. These are the seed leaves whose origins were contained within the seed. If, however, you look at the seedling of a grass or a tulip, you will notice that there is only one seed leaf. The 'dicots' have net-veined leaves while the 'monocots' have parallel-veined leaves; the latter group includes the grasses, sedges, orchids and lilies, whereas the 'dicots' contain most other plants including most trees.

2 · THE STRUCTURE OF PLANTS

CELLS

The basic building block of all plants, and indeed of all forms of life, is the cell. The very simplest plants such as bacteria and some algae consist, as we have seen, of only one cell; other algae consist of small groups of similar cells in filaments, spheres or other shapes; but the vast majority of plants are made up of literally millions of cells, many of them specialised for different functions, all working together as one unit or organism.

Despite this specialisation, the basic structure of the plant cell remains fundamentally the same with only minor modifications for different purposes. Although cells vary widely in size, they are all exceedingly small and it is very rare that they are large enough to be seen with the naked eye. An average cell is about 0.01 millimetres in diameter which means that about 10,000 could fit on the head of an old-fashioned metal pin!

Each cell consists of a cell wall made of cellulose (a compound rather like starch) which is elastic and somewhat permeable to water. Held within the cell wall is a clear viscous liquid known as cytoplasm, made up of proteins, carbohydrates, water and salts; and somewhere within the cytoplasm, often towards one edge, lies the powerhouse or 'nerve centre' of the cell—the nucleus. Under normal circumstances every cell has a nucleus, and it is the nucleus that controls the activities of the cell and, collectively, of the whole plant, dictating the final form of each cell and when, or whether, it divides to form a new cell. At the heart of this control mechanism lie the chromosomes which are contained within the nucleus. The whole process of cell division which is an essential part of plant growth is considered under 'Growth' in Chapter 3.

As the cell ages, the cytoplasm begins to develop clear fluid-filled bubbles,

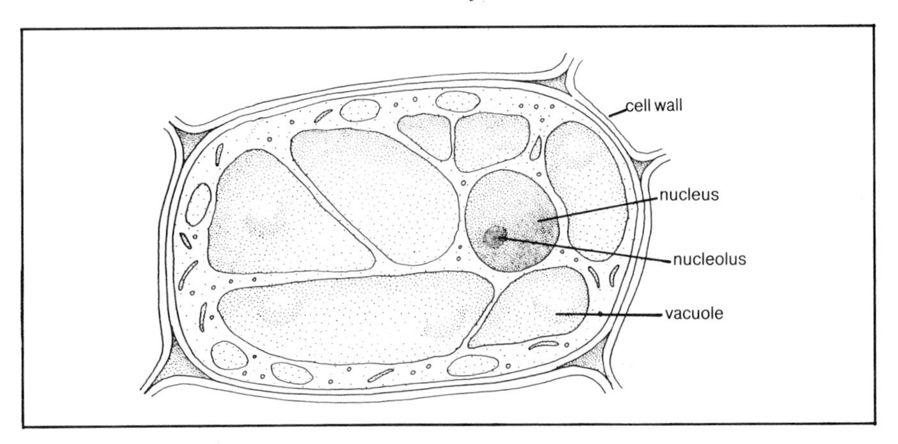

Structure of a generalised plant cell.

known as vacuoles, which contain a mixture of foods and waste products dissolved in water, varying according to the age and type of cell, the species, and the environment the plant is growing in. Gradually the vacuoles fill the cell, leaving the cytoplasm and nucleus confined to the edges of the cell. In some cells, where additional strength and rigidity is required, the cellulose cell wall may be reinforced with a hard substance called lignin (which makes up a high proportion of wood). This strengthens the cell, but also makes the cell wall less permeable and may eventually prevent it from being able to divide.

In a single-celled plant, everything that the plant needs is contained within the one cell. In more complex plants, such as flowering plants, millions of cells are joined together to make up the whole plant body, and groups of cells become specialised to form tissues to do particular jobs. In multicellular organisms, the cells are joined together with pectin (notable also for causing problems for wine-makers by preventing wine from clearing). So, if you look at a familiar plant like a dandelion, or even an oak tree, you are actually looking at millions and millions of tiny cells all working together as one unit.

ROOTS

Roots are, perhaps, the least familiar part of the plant, yet they may be much larger in extent than the above-ground parts that we see, and their functions are vital. Roots have two main functions: they anchor the plant to the ground and help it to keep upright, and they provide the rest of the plant with water and mineral salts absorbed from the soil. Roots can also act as food storage organs, such as the tap roots of familiar vegetables like carrot or parsnip, or the starchy root tubers of the early-purple orchid *Orchis mascula*. At times, they serve other functions, such as the roots of mangrove trees growing in swamps which protrude from the water to absorb oxygen, or the adhesive roots of virginia creeper (*Parthenocissus*) or ivy (*Hedera helix*) which help the plants to climb walls, trees or stones.

Two types of root systems can be distinguished: fibrous root systems and tap roots. In a fibrous root system, there is no clear dominant root, but instead there is a mass of smaller roots extending in all directions, spreading out from the base of the stem. The length and surface area of a fibrous root system developed in good conditions is astonishing: one four-month-old rye plant had nearly 14 million roots with a total surface area of 2,500 square feet (232 square metres); in addition it had some 14 billion living root hairs (small hairs with few cells that push out from the root into the soil) with a further surface area of about 4,300 square feet (400 square metres). So, this four-month-old plant had living absorbing tissue in direct contact with over 6,800 square feet (632 square metres) of soil! In another example, a two-year-old plant of couch grass (*Agropyron repens*) grown under ideal conditions, had a total root length of over 300 miles (483 kilometres)! These figures, spectacular as they are, give an idea of the importance of roots to plants.

Although grasses and cereals are amongst the best-known examples of plants with fibrous roots, many other plants possess the same system. Gardeners, for instance, will have noticed how much easier chickweed or groundsel are to pull up, with their fibrous root systems, than dandelions or docks with their solid tap roots. Plants with tap roots have one main, well-defined root, which penetrates well into the soil, from which other branch roots arise. The main tap root is frequently used for storage of food materials, and extreme cases, such as beetroots or radishes, are used as food by us.

'Adventitious' roots are another type of root. These are additional roots produced from the stem, sometimes where the main roots are waterlogged, or—as in the case of ivy—to aid the stem's adhesion to walls or trees. Their structure is similar to ordinary roots.

Exactly how the root system develops depends on the conditions it is growing under. The soil is composed of numerous tiny particles of silt, sand, clay and organic matter, with mineral-rich water trapped between the particles. The roots grow and penetrate between the particles to absorb the water and salts. In soft, damp, easy soil the roots are densely packed in a compact ball. In dry conditions, roots may extend for a very long distance,

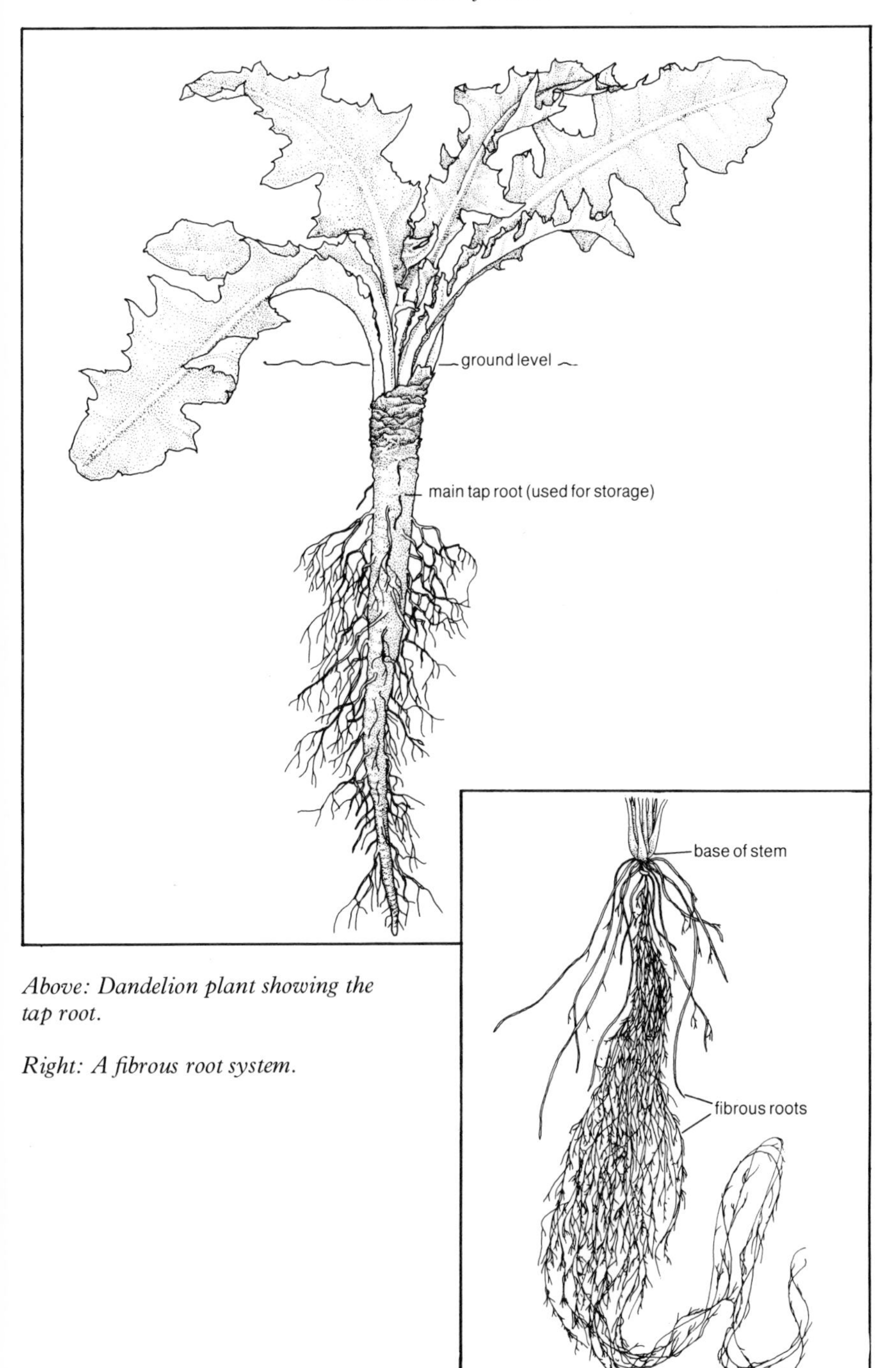

Above: Dandelion plant showing the tap root.

Right: A fibrous root system.

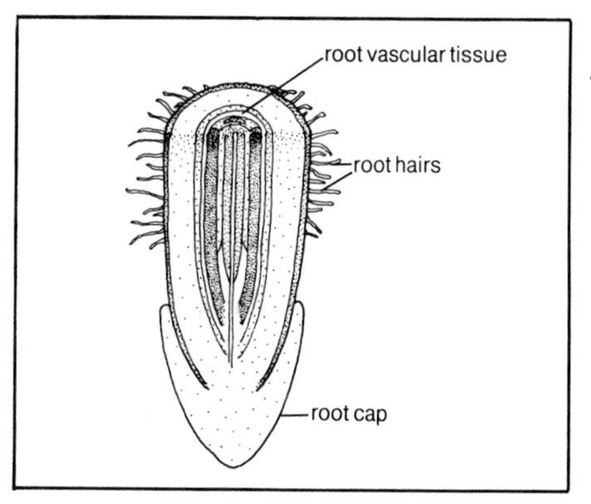

Section through the tip of a growing root.

seeking out sources of water (because, as we shall see in Chapter 3, some roots grow towards water). There are many examples of plants growing in arid conditions where a tiny stunted above-ground plant supports an enormous tap root system below ground.

If we look more closely at a root under magnification, we can see that it has a definite structure, most clearly seen in a longitudinal section. At the extreme tip of the root is a protective cap of cells which have ceased to grow or divide; this is known as the root cap. Immediately behind the root cap, and protected by it as the root pushes through the soil, is a group of actively-growing and dividing cells known as the root meristem. As these cells elongate and split continuously, so they push the root cap further into the soil and the whole root grows. Behind this area of active growth, away from the moving part of the root, is an area in which the tiny root hairs push out from the outer layers of the root into the soil, helping to expand the plant's capability to absorb water and salts. The root hairs are the main points through which water is absorbed into the plant, and they are constantly being replaced as the old ones are abraded.

Around the roots and root hairs of many plants, especially amongst the orchids, heathers and some forest trees, there is a cobweb-like net of fungal strands. These fungal strands, or hyphae, form a special relationship with the plant which, because both partners appear to benefit from it, is known as a symbiotic relationship. It is almost as if the fungus tries to invade the roots but is only partially successful; the attack is contained and the roots absorb

The roots of plants produce numerous fine root hairs which penetrate the soil and absorb water. They are clearly visible here in this radish seedling germinated on damp paper.

water and mineral salts from the fungi while they, in turn, derive some nourishment from the roots. This peculiar specialised relationship between fungi and flowering plants is known as a mycorrhizal association, and without the help of the fungus many orchids could not survive.

So, in one way or another, the roots absorb water. The main point at which the water is required, however, is in the leaves and, to a lesser extent, the stem. At the same time, the roots require food to sustain their growth and maintenance; the material to produce 14 million roots in four months has to come from somewhere. This interchange is achieved by a specialised system of transport veins which run from near the tip of the roots, up the stem, to the leaves (where they become visible as 'veins'). This transport system consists, in fact, of two quite distinct elements—the xylem and the phloem.

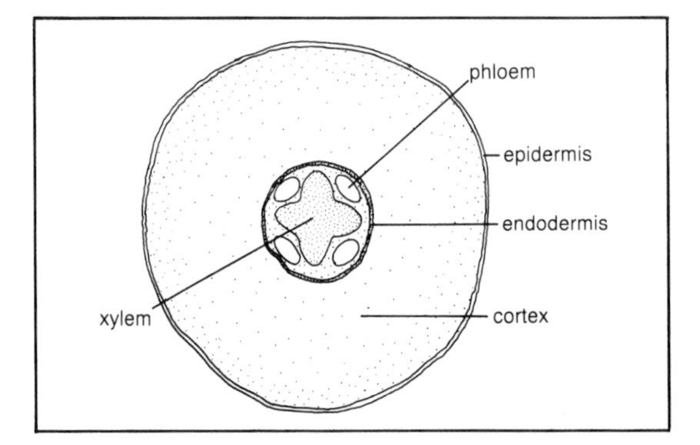

Cross-section of a root.

The xylem consists of strongly lignified but dead cells, most of whose end walls have broken down to form almost continuous tubes running up through the plant. These vessels carry water, and any dissolved salts, up through the plant. The phloem is quite distinct and consists solely of living cells whose function is to transport food materials around the plant in whatever direction they are needed. They consist of tubular, perforated cells, known as sieve tubes and laid end to end, and adjacent companion cells (for want of a better term) which appear to play a part in regulating the rate of flow of food through the sieve tubes. The phloem and the xylem together are known as the vascular system of the plant (rather like the blood system of mammals), and in addition to their transport function they play a part in increasing the rigidity of the plant. In the root, the vascular system forms a central core, with the woody xylem in the centre and the phloem scattered around it. In stems, however, the arrangement is different, and separate little bundles of xylem and phloem together occur scattered around the

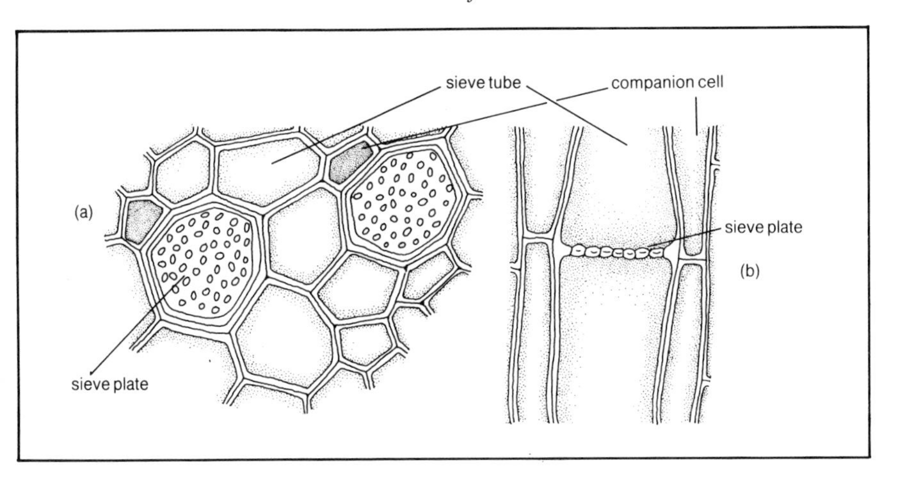

Phloem (food-conducting) cells in (a) transverse and (b) longitudinal section.

circumference of the stem (with the exception of genuinely woody plants, as described below). This change from a central core to peripheral bundles appears to relate to the support needed to counter the different stresses encountered by an aerial stem and a subterranean root.

STEMS

The essential function of a stem is to hold the leaves and flowers in their most appropriate positions and to connect them to their water supply, the roots. Almost all stems do this, and almost all flowering plants have stems—but what a difference in the way they do it! The trunk of a 250-foot (76-metre) high giant redwood is a stem, yet so is the tiny fragile thread that connects the floating leaves of a duckweed (*Lemna*) to its roots. Some stems are rigid and hold their leaves and flowers in a fixed position, whereas others are designed for twining or scrambling up other vegetation, such as the runner bean, or the black bryony (*Tamus communis*). Yet, despite this great variety of form, all stems, with the exception of truly woody ones (and even these start off in the same mould) have a similar basic structure.

If you look at a plant of the familiar herbaceous weed chickweed (*Stellaria media*), the general external structure of the stem is readily visible. It is green throughout, as many stems are, except in the oldest portions. The stem emerges from the ground as a straight, flexible thin cylinder, though after a while a point is reached where the stem thickens slightly, and a pair of leaves (one on either side) sprouts from the stem. Beyond this the stem

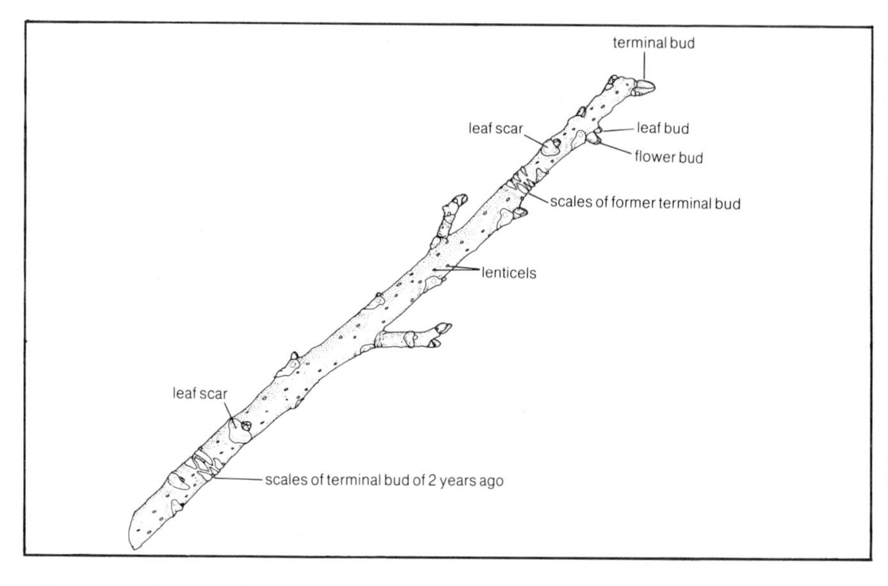

Structure of a typical woody twig (walnut).

continues as a straight cylinder, perhaps slightly thinner than before, though in chickweed two stems often arise from this point where the leaves are borne. The points where the stem is thickened and the leaves are borne is known as the node, and the lengths of stem between the nodes are known as internodes. Progressing up the stem, there is a series of these nodes with leaves, and progressively shorter and thinner internodes between them. Increasing numbers of flowers emerge from the bases of the leaves at the nodes, until eventually the stem appears to end in a pair of leaves. Closer examination indicates however that these leaves contain one or more buds between them, ready to burst into action, and this pattern of growth can go on indefinitely. Each main stem has a terminal growing point, and immediately below this, in the angle between the highest leaf and the stem (the leaf axil) is another bud. This bud may grow into a flower, another stem, or it may simply remain dormant as long as the main terminal bud keeps growing. If the terminal bud is damaged or removed, this lateral bud will spring into growth (*see* Chapter 3), and take over the function of the terminal bud.

The internal structure of the stem closely reflects the functions that it has to perform. The inside of a bud reveals that it is like a condensed miniature shoot with preformed nodes and internodes with little space between them; if you can imagine a cross-section of an onion, reduced to the size of a bud,

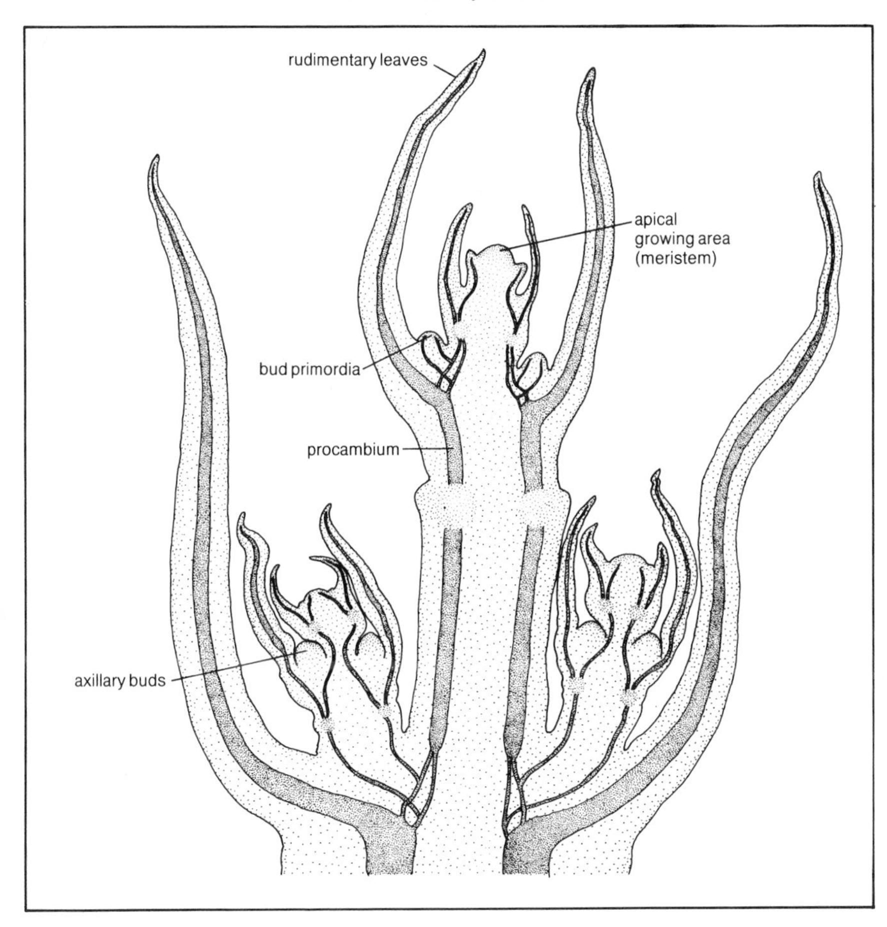

Section of leaf bud of lilac (approximately × 25).

this will give a good picture of what a bud consists of. The central portion, the tip, contains the growing point or meristem, and in an actively growing shoot it is here that the greatest growth is taking place. Unlike the roots, there is no need for a hard protective cap as the stem is not being constantly abraded by soil particles.

The internal structure of the stem is also surprisingly constant, with minor variations. In cross-section, there is a clearly-defined outer layer of slightly thickened cells forming a protective impermeable skin, known as the epidermis. The roots have no such skin, as one of their functions is to absorb water; in contrast the stems need to prevent unnecessary water loss so that the vital liquid can be used where it is needed. Immediately below

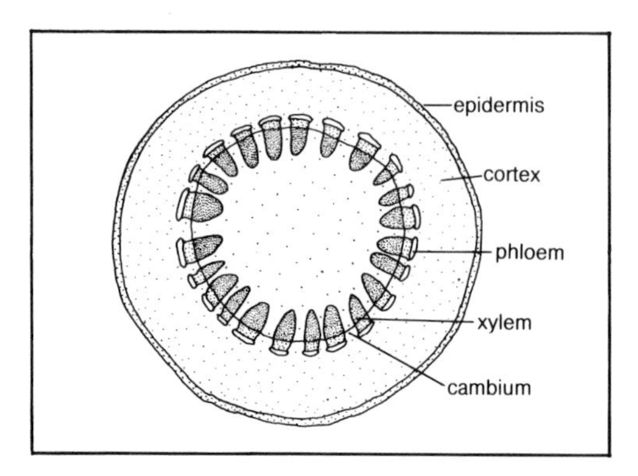

Cross-section of a young woody stem.

this, there is often a ring or series of cords of flexible strengthening tissue; in some plants such as the white deadnettle (*Lamium album*) this is visible as a series of ridges running up the stem. Just inside these threads, if they occur, lies the vascular conducting tissue—the xylem and the phloem—which instead of occurring in a central core as it does in the root has spread out into a series of peripheral 'bundles', each containing some water-transporting tissue (xylem) and food-conducting tissue (phloem). The central portion of the stem is generally occupied by undifferentiated 'ground' tissue, correctly known as parenchyma, and although the cells are unthickened and unsupported they give the stem its rigidity entirely through the water pressure within them, rather like an inflated balloon; if the cells lose their turgor, the stem wilts and becomes less rigid. So, the stem is rather like an insulated cable with an outer skin, some layers or threads of supporting tissue below the skin, and a series of conducting vessels running through it, all branching out to leaves or flowers along the way.

A close look at a cut log or felled tree will quickly show that the structure of a woody stem is not quite like that. Although the young stems of woody plants may start off looking like the stems described above, they contain within them the elements of a quite different system of growth which eventually takes over and alters the whole character of the stem. The process itself is known as secondary thickening, and the way in which it works is described in more detail under 'Growth' in Chapter 3. In completely woody plants, such as an oak tree, the stem or trunk consists of an outer corky protective bark, a thin layer of phloem and associated tissues just below the bark, and then the great mass of the tree, from here to the centre (i.e. the wood), is composed of heavily lignified xylem vessels laid down in annual rings. Here and there a 'ray' of unlignified tissue runs across the trunk from the centre to the edge, and there are occasional 'knots' of xylem and phloem

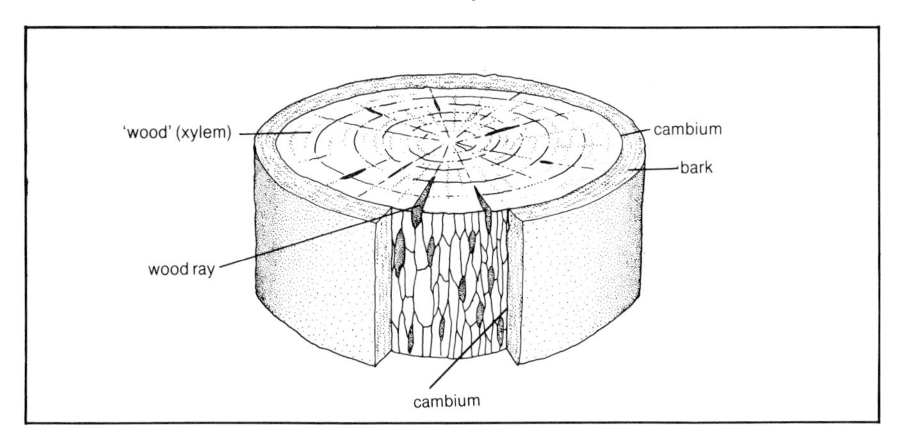

Cross-section of a woody stem.

running at an angle to the remainder, where a woody branch begins to emerge from the main trunk. (This is one reason why timber trees are often grown in dense plantations—the absence of light at the lower levels means that branches do not arise from the main stem, and so a knot-free timber is obtained.)

Modified stems

Besides their characteristic functions of support and transport, stems may become modified in some plants to fulfil other functions. In many cacti, for example, the stem has become greatly modified to form an enormous water-storage organ, whose surface layers are also able to produce food by photo-synthesis (*see* Chapter 3); the leaves are completely reduced to spines and serve a subordinate protective function.

Some forms of tuber are actually modified stems, the best-known example being the potato. Sections of the stem enlarge to form food storage organs (hence their nutritive value for humans) which, although well below ground, betray their origins by the buds on them (the 'eyes') and their ability to go green in the light.

Some stems grow horizontally, often below ground. Where they grow below the ground, they are known as rhizomes, though they still retain the characteristic stem structure with nodes and internodes; from these nodes adventitious roots grow downwards, and shoots grow upwards towards the light. The 'roots' of the dreaded couch grass (*Agropyron repens*) are actually rhizomes sending up shoots wherever they go. Some rhizomes fulfil a storage function, such as the swollen rhizomes of the iris.

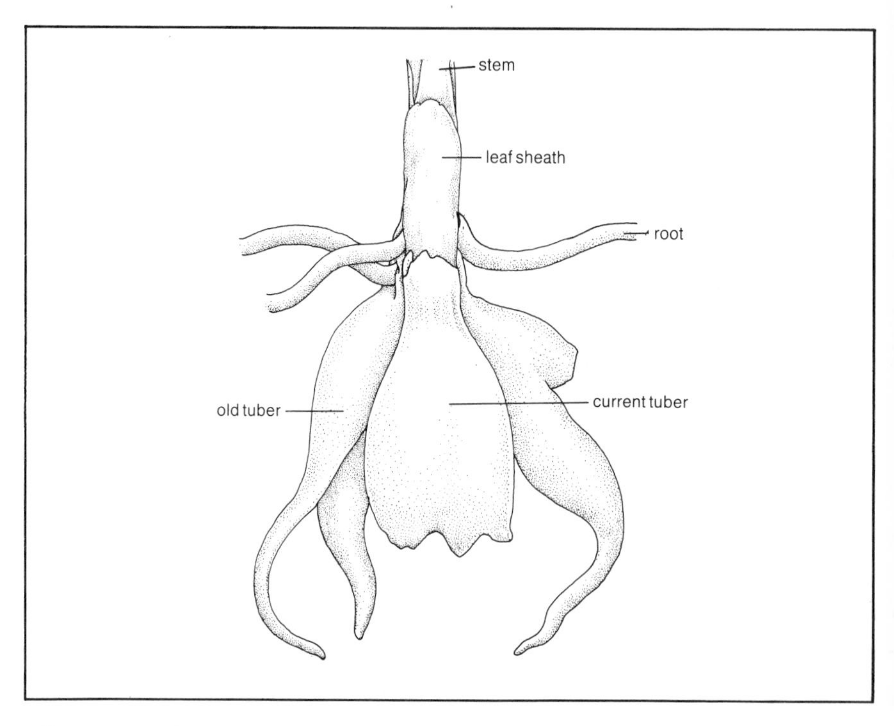

*Root system of spotted orchid (*Dactylorhiza*).*

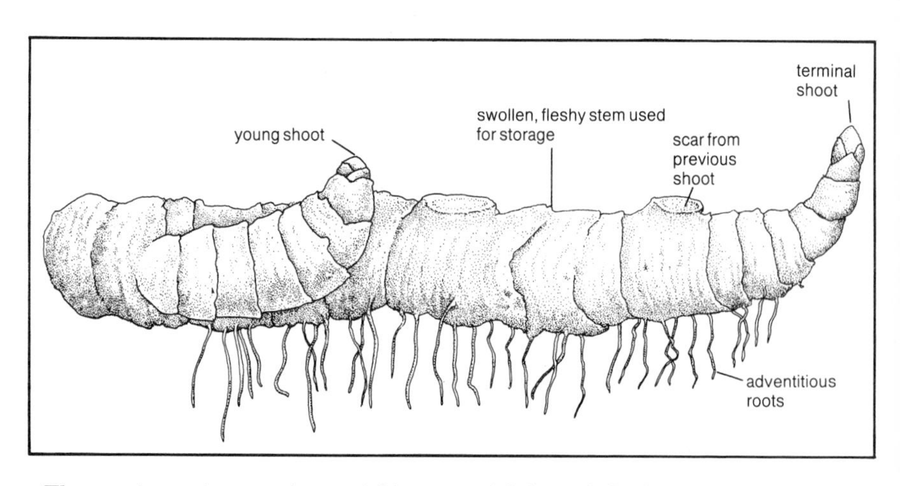

The creeping underground stems (rhizomes) of Solomon's Seal.

*The flowers of butcher's broom (*Ruscus aculeatus*) are produced very early in the year on modified stems that look like leaves.*

Where the stems grow horizontally but above ground they are known as creeping stems when the main stem grows horizontally, e.g. in creeping jenny (*Lysimachia nummularia*), and as runners or stolons where the horizontal stems develop from axillary buds, e.g. the strawberry with its familiar runners developing from an upright main stem. This growth habit gives the plant the ability to spread quickly and securely in a favourable place, and many of the most persistent garden weeds have horizontal stems.

Some tendrils are formed from modified stems (though many are modified leaves) and some thorns are modified whole shoots, e.g. in the hawthorn (*Crataegus monogyna*) whereas others such as rose thorns are outgrowths from the stem. The familiar corms of crocuses and other plants are stems modified to do the job of food storage for overwintering as well as producing new plants. As the reserves of the old corm are exhausted a new one is produced above it, which is then pulled down by contractile roots to the level of the old one. New plants are produced as secondary corms to one side.

One peculiar modification of the stem is found in the butcher's broom plant (*Ruscus aculeatus*). The whole plant apparently consists of a tight mass of spiny grazing-resistant leaves; but close examination of the flowers in February or the large red berries reveals that they are actually attached to the middle of these apparent leaves. In fact the 'leaves' are modified stems, and the true leaves are reduced to tiny scales at the base of the flowers.

LEAVES

The leaves of a plant are the food-producing organs, using the energy from sunlight to convert water and carbon dioxide from the air into sugars by the process known as photosynthesis (*see* Chapter 3). Their structure and arrangement reflects the need for them to obtain as much sunlight as possible and to avoid cutting too much out from leaves below them on the same plant. A secondary function is to act as waste-disposal organs, accumulating unwanted materials before finally dropping off, taking the waste materials with them.

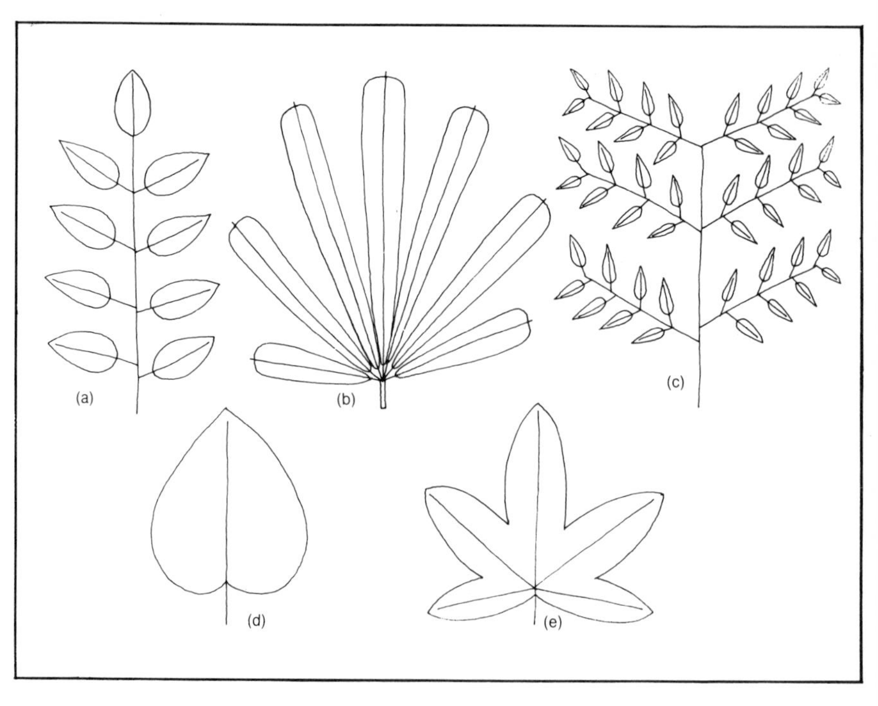

Some examples of different leaf shapes: (a) pinnate; (b) palmate; (c) bipinnate; (d) cordate; (e) palmatifid.

Leaf shapes

The shape of leaves varies enormously, and it is obvious that no one leaf shape solves all the problems, even for plants growing under apparently similar circumstances. Perhaps the commonest shape is roughly ovate, like the leaves of a beech tree; some leaves are divided into finger-like lobes from the base, like a horse-chestnut leaf, and these are known as palmate leaves; others are divided into lobes or leaflets attached along the main vein of the leaf, like the leaves of mountain ash (*Sorbus aucuparia*) or Jacob's ladder (*Polemonium caeruleum*). The latter are called pinnate leaves, and where the leaflets are divided again to their central vein as in the male fern (*Dryopteris filix-mas*), for example, they are known as bipinnate. There are numerous variations on these themes, which, when coupled with such variable characters as toothed margins, hairiness, length of leaf stalk and so on, mean that almost every species has a leaf that is unique.

Most leaves have a stalk, known as the petiole, attaching it to the plant stem, though others are attached directly without a stalk and are known as sessile leaves. The petiole carries the vascular tissue (the xylem and phloem)

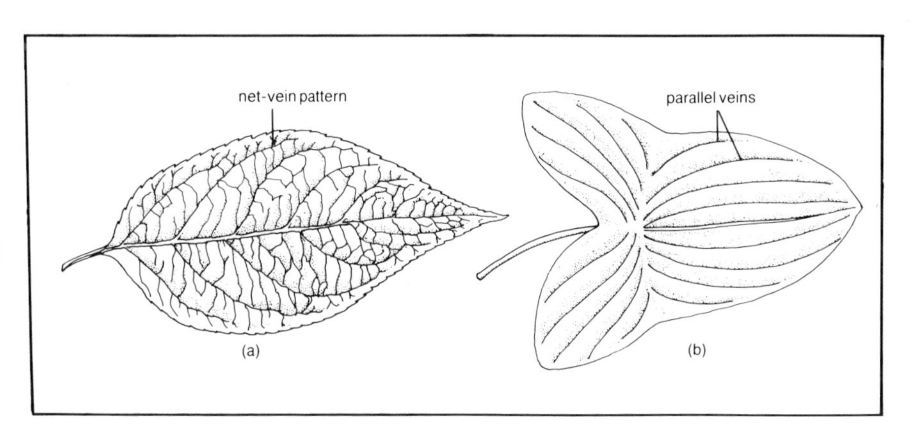

Leaf-shape and vein pattern of (a) a typical dicotyledonous plant and (b) a monocotyledonous plant.

from the main stem which branches out into the veins that are visible in most leaves. The pattern of veins varies widely but the most consistent difference is between the dicotyledons and the monocotyledons (*see* Chapter 1). The leaves of dicotyledons have a network of veins spreading out through the leaf like the tributaries of a river; monocotyledonous leaves have parallel veins running the length of the leaf, like those of the gladiolus or lily of the valley. There are very few exceptions to this rule, and it is one of the easiest ways of distinguishing the two groups of plants. The function of the veins is to bring water and mineral salts to the cells of the leaf (through the xylem vessels) and to distribute the foodstuffs manufactured by the leaf to the rest of the plant (through the phloem cells).

Leaf structure

The outermost 'skin' of the leaf consists of a single layer of cells, tightly packed together, known as the epidermis. The upper skin has a waterproof covering of a fatty substance called cutin covered by a thin waxy layer, to prevent water loss from the leaf. The lower epidermis has a much less well-marked layer of cutin as it is less exposed to the winds and sun. This is clearly visible in many leaves where the upper surface is dark and glossy but the lower surface is pale and matt.

Between the upper and lower layers of epidermis lies the main mass of the leaf, known as the mesophyll. The upper layer of the mesophyll consists of tall thin cells with thin walls and air spaces between them—these are known, because of their appearance under the microscope, as the palisade cells. They all contain masses of small globular green bodies known as chloroplasts, and it is in these that the process of photosynthesis goes on. The chloroplasts also give the leaf its green colour; although there are no

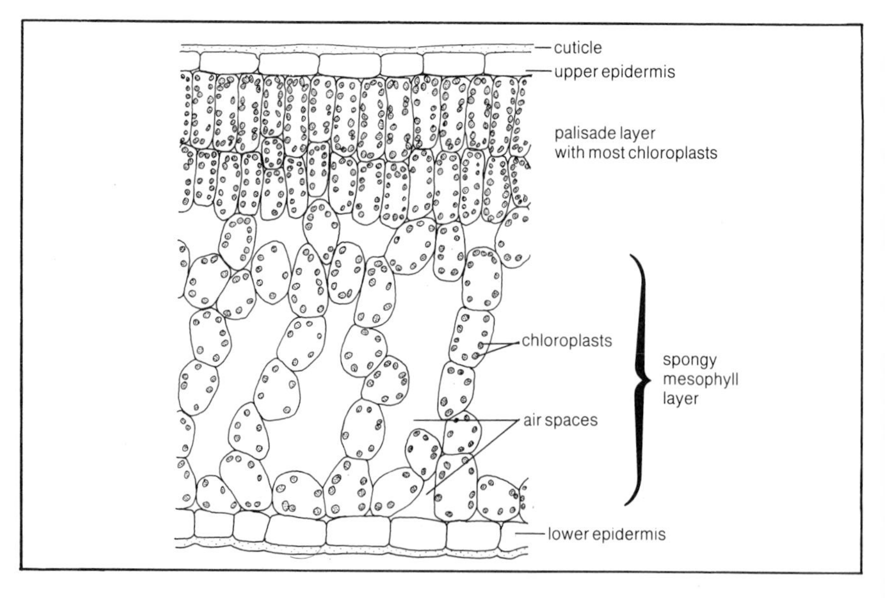

Cross-section of a leaf showing the different cell layers.

*The real flower of painter's palette (*Anthurium andreanum*) is the white central spike, but the bright pink modified leaf gives it its colour and attractive form.*

chloroplasts in the epidermis the colour of the lower layers shows through. Below the palisade layer is a layer of more rounded cells with large air spaces between them and fewer chloroplasts. This is the spongy mesophyll, which also accounts for some photosynthesis, though less than the palisade layer as less light reaches through to it.

The process of photosynthesis needs air to keep it going, and the waste gases from photosynthesis and respiration have to be disposed of; yet, at the same time, excessive loss of water vapour has to be prevented or the plant will wilt. To achieve the right balance, a system of adjustable pores has developed which can close when water is in short supply or when photosynthesis is not proceeding (i.e. in the dark). These pores are known as stomata, and in most leaves they are concentrated on the lower surface of the leaf where water loss is lowest. The structure of the stomata is more or less constant; each consists of a pore surrounded by two sausage-shaped cells known as the guard cells (which, unlike the rest of the cells in the epidermis,

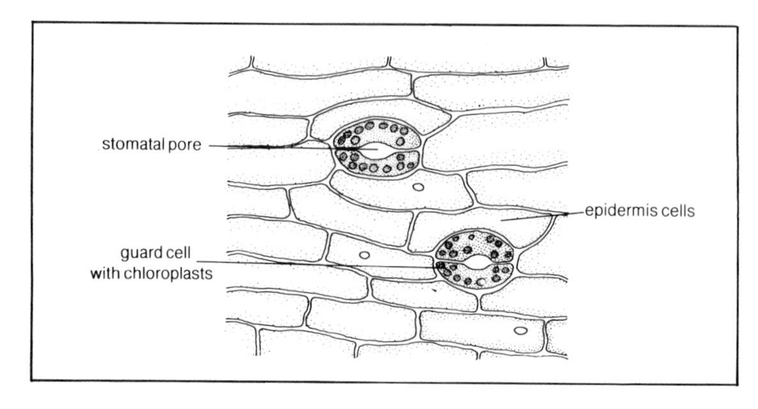

View of stomata in lower leaf surface.

contain chloroplasts). The part of the cell wall next to the pore is thickened and therefore less flexible, whereas the outer wall is unthickened and more flexible. If conditions are moist, the cell absorbs water and becomes turgid (i.e. at full pressure). The outer walls expand most so the cell becomes kidney-shaped and the central pore opens. When conditions are dry, the cells become flaccid and straight, and the pore closes. This means, of course, that the plant's capability for photosynthesis is greatly reduced when conditions are very dry, and this is one reason why plants grow more slowly in dry conditions. The pores are extremely small, but they are present in large numbers; for example, the common bird's-eye speedwell (*Veronica chamaedrys*) has about 230 stomata per square millimetre (about the size of a small pinhead) on the under-surface of the leaf, and about 40 per square millimetre on the upper surface, though this varies enormously between species. Monocotyledonous plants tend to have similar numbers on each surface; for example, barley has about 50 per square millimetre on each surface, whereas aquatic plants with floating leaves have, not surprisingly, many more stomata on the upper surface. The similarity of structure and pattern of the stomata in almost all plants is remarkable, suggesting that it is the best available solution to the problems that plants face.

Leaf modifications

Although, as we have seen, the basic function of leaves is as food-producing organs, in some plants they may become modified to perform quite different functions. Perhaps the commonest modification is where a whole leaf, or part of a leaf, has become modified into a tendril to help the plant to climb; a familiar example is the garden pea (*Pisum sativum*) and some vetches. In

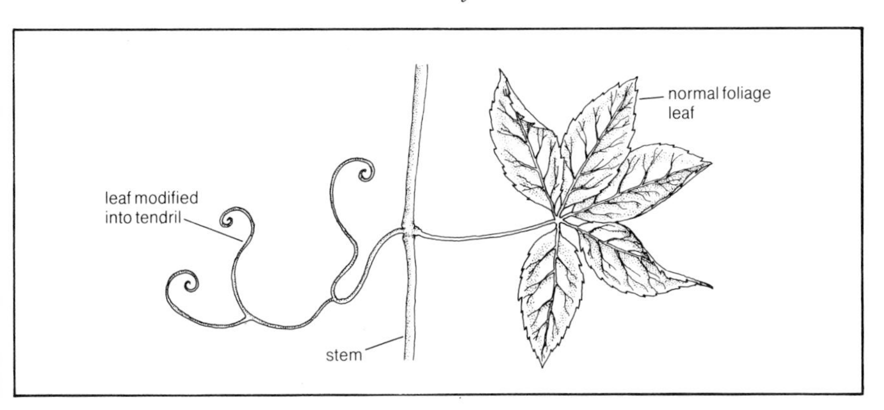

Leaf modified into tendril.

other plants, the leaves form an important part of the flower system; the flower itself may be inconspicuous but the leaf is modified into a colourful 'bract' which forms the most visible and attractive part of the flower, e.g. the red 'flowers' of poinsettia (*Euphorbia pulcherrima*) are actually bracts, and the same is true of painter's palette (*Anthurium*), bougainvilleas and the attractive alpine plant dwarf cornel (*Cornus suecica*).

In other cases, the leaf is modified to serve a protective function, e.g. the spines on the leaves of the holly (*Ilex aquifolium*), or in barberry (*Berberis*) where the whole leaf may be modified into a spine. The leaves of stinging nettles have tiny single-celled hairs extending out from the epidermis, each

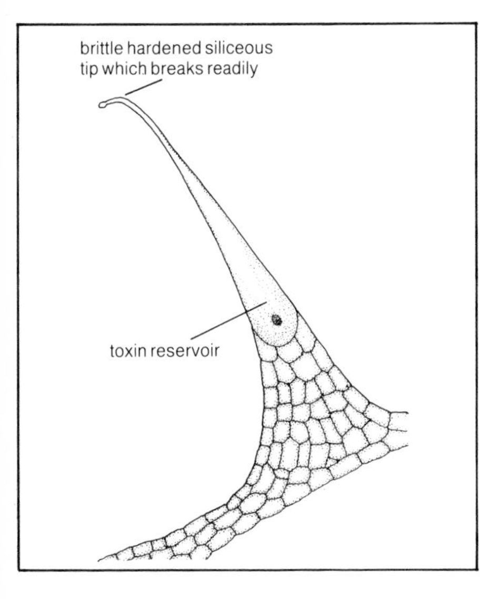

*Diagram of stinging hair of stinging nettle (*Urtica dioica*).*

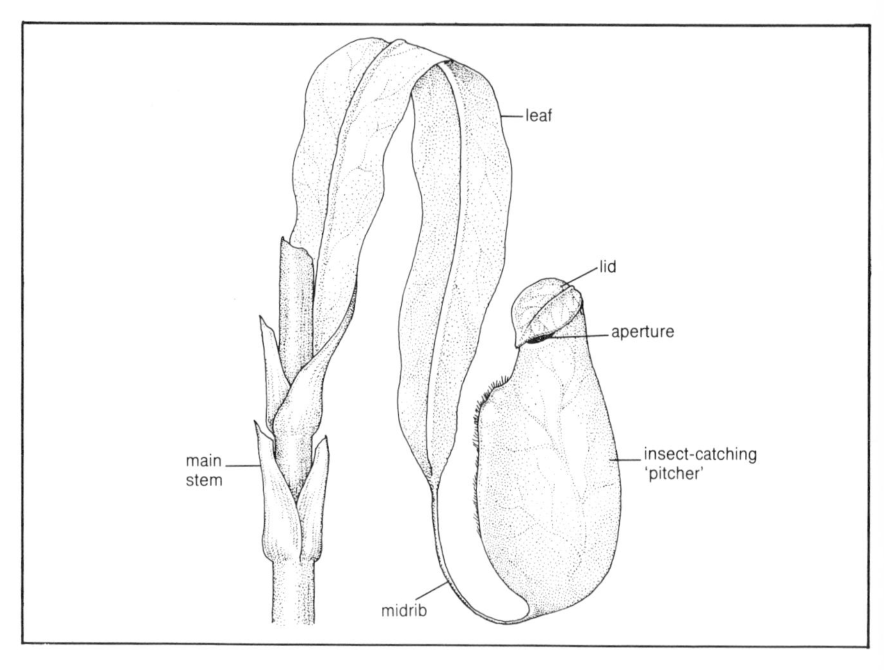

*The leaf of the pitcher plant (*Nepenthes*) is developed into a pitcher which contains water. Insects fall into the water, drown and are digested by the plant.*

of which has a fragile tip and a bulb at the base containing a complex toxic protein. The effects of this system are all too well known!

Perhaps the most dramatic of the leaf adaptations are those designed to supplement the food supply by catching insects and digesting them. The leaves of sundew (*Drosera*) or butterwort (*Pinguicula*) still look and perform like leaves despite their sticky hairs, whereas the leaves of the venus fly trap (*Dionaea muscipula*) are more modified with their ability to close rapidly over an unfortunate insect. The least leaf-like adaptation, but a leaf nevertheless, is the insect-catching pitcher of pitcher plants (*Sarracenia* and *Nepenthes*). The process of insect-catching by plants and the reasons why it has developed are considered in more detail in Chapter 3.

In summary, leaves are normally the basic food-producing organs but they are frequently modified to perform other functions as well as, or occasionally instead of, their main one.

3 · THE DAY-TO-DAY LIFE OF PLANTS

Although it is not always readily apparent, plants grow, absorb water through the roots, manufacture food, 'breathe', react to light and dark and cold and heat. All these processes go on in plants, controlled both by the changing external environment and by an internal control system.

PHOTOSYNTHESIS

The most important and significant process that goes on in plants, and indeed the basis of almost all life on earth, is photosynthesis. In simple terms, this is the process in which plants absorb carbon dioxide from the air and combine it with water in the presence of sunlight to produce sugars with the release of oxygen. The site of this remarkable reaction is in the green chloroplasts of leaves and stems, acting rather like miniature solar panels capturing and storing the sun's energy far more efficiently than anything yet devised by man.

However, the process is not a simple one, and it is worth examining in a little more detail. The air that we breathe consists of several different gases, notably nitrogen (80 per cent) which is almost inert (but *see* p. 52), oxygen (almost 20 per cent), vital to most forms of animal life, and carbon dioxide (0.03 per cent), together with a few other less significant gases in small amounts. Air enters the leaf by the system of interconnecting air spaces already described. Thus it reaches the outer surface of the palisade cells which are damp and able to absorb the carbon dioxide from the air mixture. The carbon dioxide passes through the cytoplasm to reach the chloroplasts. The chloroplasts themselves are mobile within the cell and tend to cluster at whatever part of the cell is closest to light. When the light reaching the leaf is above a certain intensity (and this level varies according to the species, depending on the light levels they are adapted to) the green pigment in the chloroplasts, chlorophyll, emits a tiny electric current, which is 'harvested' by special electron-carrying molecules. This electric current is used as the energy source to split water (H_2O) into its component molecules of hydrogen and oxygen and to form an energy-rich compound known as ATP (adenosine triphosphate) which can then be used as a source of energy when required. The oxygen plays no further part in photosynthesis, and any surplus is emitted as oxygen gas into the atmosphere, though some may be consumed by the plant in the process of respiration (*see* below). On balance, in a healthy plant, more oxygen is given off than carbon dioxide consumed, and plants are an essential part of the system that produces the atmosphere we are able to breathe.

The 'free' hydrogen is then combined with the carbon dioxide to produce glucose in a highly complex series of reactions using the energy from ATP, which can take place in the dark. The glucose is the basic food of the plant and it can be converted into other foods as required, or broken down to provide energy in the process of respiration. For example, only plants can synthesise proteins from minerals and sugars alone—animals are unable to do this.

Only a proportion of the light reaching the leaf actually reaches the site of activity in the chloroplast, and the chlorophyll is only able to use a proportion of this light since some wavelengths are unsuitable. So it can be seen how vital light, water and carbon dioxide are to the plant, as all three must be present together in adequate quantities before the plant can carry out its most basic process.

THE NEED FOR NUTRIENTS

Even though the basic ingredients of carbon dioxide, water and sunlight may be present in abundance, it is well known that plants grow better on some soils than on others. Similarly, the addition of fertilisers will dramatically affect the growth of plants on 'poor' soil. Why should this be so when the plant apparently has all the ingredients necessary for photosynthesis? The answer is both complex and simple. The living plant is made up of enormous numbers of chemical compounds—enzymes, hormones, pigments (like chlorophyll), waste products, foods, proteins—and many of these compounds contain simple inorganic substances derived from salts, such as nitrates, phosphates and potassium. For example, phosphates and sulphates form part of many proteins, potassium is essential for cell division, among

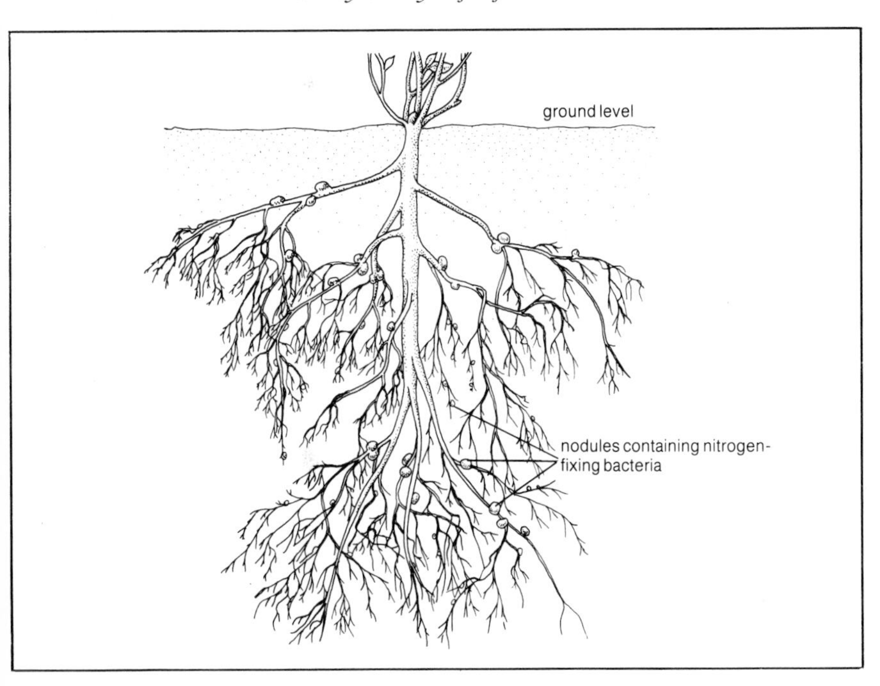

Bacterial root nodules on the roots of a leguminous plant.

other things, and magnesium is an essential part of chlorophyll. There are six elements that are widely used in plants, the so-called macronutrient elements, such as calcium, magnesium, phosphorus and so on, and many other trace elements, such as boron or copper which are known to be essential to plants but which need only be present in very small amounts. Generally, these elements are dissolved in some form or other in the soil water, but most soils are deficient in one or other of the nutrients. Wild plants have adapted to this, and different species or varieties grow on different soils with their differing nutrient levels. With crops, however, maximum growth is required from the plants whatever the soil, and fertilisers are used to rectify any imbalances, especially in the critical three elements—nitrogen, phosphorus and potassium. Plants will exhibit signs of deficiency if a particular element is lacking; for example, plants growing without adequate magnesium are yellowish (chlorotic) and weak because they cannot form enough chlorophyll. The situation may be complicated because some nutrients, if

*Overleaf: The leaves of this long-leaved sundew (*Drosera intermedia*) are strongly modified to allow them to trap and digest insects by means of sticky glandular hairs.*

present in excess, prevent the uptake of others; for example, the excess calcium in lime soils may prevent the uptake of iron, though this can be partly overcome where necessary by the addition of substances called 'chelating agents' to the soil, which allow the iron to be taken up despite the calcium.

Some plants have evolved special additional ways of acquiring nutrients. The two main ways are through nitrogen fixation and by insectivory (insect-eating). Nitrogen is a vital part of proteins, chlorophyll, hormones and many other substances, so it is needed in considerable quantities which may not always be available in the soil. Although nitrogen forms such a high proportion of the air, this gaseous nitrogen cannot normally be used by plants since the roots can only absorb the dissolved salts of nitrogen such as nitrates. Some bacteria, however, can use nitrogen directly from the air and 'fix' it to form nitrates and other nitrogen compounds. Some plants, and especially the legumes (peas, beans, etc.) have developed a special relationship with nitrogen-fixing bacteria in which the bacteria live in nodules on the root of the plant and there, adequately supplied with the other nutrients they require, they fix large quantities of nitrogen, much of which becomes available to the host plant. Such a mutually beneficial relationship is known as symbiosis.

Plants growing in very acid or waterlogged environments may have considerable difficulties in obtaining enough nitrogen, and a number of bog plants have developed methods of catching and digesting insects. All involve some method of trapping the insect, preventing it from leaving, and ultimately digesting it and absorbing the nutrients from it. For example, sundews, familiar bog plants of temperate regions worldwide, have long hairs on their leaves which secrete droplets of a highly sticky substance to which the unfortunate insects become attached. The hairs then bend over and press the insect against the leaf, from which digesting enzymes are secreted, and eventually the soluble parts of the insect are dissolved and absorbed.

PARASITES AND SAPROPHYTES

A few flowering plants have overcome the need for photosynthesis by relying entirely on the production of other plants. Such plants are parasites, and they are frequently devoid of green colouring, with their leaves (since they now serve no function) reduced to tiny colourless scales. The toothwort (*Lathraea squamaria*) for example, is a parasite on the roots of hazel and maple, depending totally on them for all its support. The plants of toothwort apparently live for a very long time, increasing in size, and in their effect on the host, throughout this period. Other parasites, such as the broomrapes (*Orobanche* spp.) grow mainly on herbaceous plants, and tend to be much shorter lived. The common dodder (*Cuscuta epithymum*) is a parasite which sprawls in a network of fine stems and pale flowers over many moorland plants.

A few flowering plants look like parasites in that they are lacking in green colour and have no developed leaves, but yet they do not depend on other plants for their nutrition. Instead they are able, often with the help of fungi in the soil, to live on the breakdown products of rotting leaves, wood and so on. The bird's-nest orchid (*Neottia nidus-avis*) is a striking example of such plants, which are known as saprophytes.

There is another group of plants which show no outward signs of differing from normal self-supporting plants; they have a full system of green leaves, yet they are able to attach themselves to the roots of other plants and remove some of the products from the host plant. They are known as hemiparasites, and they are most frequent in grassland where this parasitism gives them a competitive edge, weakening their neighbours and strengthening themselves. Hayfields, for example, are frequently dominated by the bright yellow flowers of hay-rattle (*Rhinanthus minor*), and other parasites include the eyebrights (*Euphrasia* spp.) and the louseworts (*Pedicularis* spp.).

RESPIRATION

Living organisms constantly require energy in a usable form. Many of the processes in the plant, such as growth, reproduction, the synthesis of complex molecules and so on, require energy to carry them out. The glucose produced in photosynthesis is the basic unit which provides this energy (though other more complex molecules, e.g. fats, can be broken down to glucose to provide energy). The process by which glucose is broken down to yield energy, carbon dioxide and water is known as respiration, and it is common to most living things, including animals. Essentially, it is the reverse of photosynthesis, using part of the product of photosynthesis to yield energy where it is required and to return carbon dioxide and water to the atmosphere. In other words, the sun's energy is stored in glucose which is manufactured when the sun's energy is present, and then used as and when required in the plant body.

The actual process of respiration is, typically, a complex series of reactions but the net result is that the glucose is broken down to its components of carbon dioxide and water, and a store of molecules of ATP are formed.

*The pink stems of the parasitic dodder (*Cuscuta epithymum*) grow over other plants, such as gorse, and derive all their food from their 'host' through special attachments which penetrate the cells of the host.*

This ATP is the universal currency of energy in plants and animals and when energy is required anywhere ATP is simply broken down to its precursor ADP (adenosine diphosphate) and energy is released. Respiration takes place in all living cells, whenever the temperature is high enough, and is part of the continuous metabolism of the plant, enabling it to grow and reproduce.

So, in summary, the glucose manufactured in photosynthesis can either be built up into more complex carbohydrates (e.g. cellulose), fats or proteins using energy, or broken down to yield energy. Both processes are going on all the time.

WATER UPTAKE AND TRANSPIRATION

The role of water in the plant has now become clearer. We know that it is an essential part of photosynthesis and that all cells contain water when they are alive and healthy. The structure of the roots and the transport system has already been described, but how does the water get through the plant, let alone to the topmost leaves of a tall tree?

In the soil, water is taken up by the root hairs because the concentration of salts is higher in the cells of the root hairs than in the soil water, and water moves towards the higher concentration of salts, through an appropriate membrane (in this case the cell cytoplasm), by a physical process known as osmosis. (You may have noticed that vegetables, if salted and left overnight, will be very watery by morning and this is partly caused by water coming out of the cells by osmosis, being drawn across the cell walls towards the highly-concentrated salt solution outside; it is the reverse of the process above.)

The water then passes by diffusion into the root tissues and so to the xylem vessels, the start of the continuous system that carries the water throughout the plant. At the other end of the system are the leaves with their

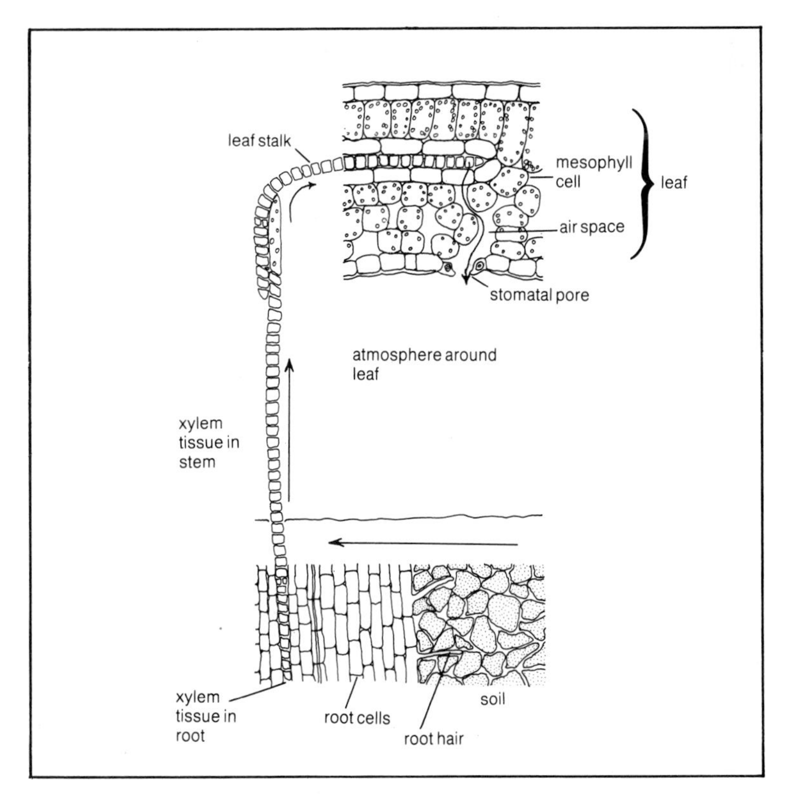

Diagrammatic representation of movement of water from the soil through the plant to the atmosphere—the 'transpiration stream'.

stomatal pores allowing water vapour to escape, and it is the force generated by this evaporation that keeps the whole system going. The process of water loss by the leaves of plants is known as transpiration, and it is estimated that 95 to 99 per cent of the water taken up by the roots is lost to the air, and only the remaining one to five per cent is actually used by the plant.

As water is lost through the stomata by evaporation, so water moves from the air spaces in the leaf to replace it; water then moves from adjacent cells or directly from the xylem vessels to replace this water, and so a continuous process which exerts a very considerable pull is set up, with the result that water continually moves from the dilute solutions of the soil, through the slightly more concentrated cell solutions in the root, and eventually to the cells of the leaf.

Although it is clearly a problem for the plant to have continually to absorb water to the extent of 100 times more than it requires for its own direct uses, the whole process of transpiration is clearly a necessary one allowing water and dissolved minerals to reach all parts of the plant. If plants are grown in conditions of 100 per cent humidity, where evaporation is prevented, they may become distorted and unhealthy, thus indicating the value of this continuous water stream.

When plants are growing in conditions of high air humidity, but with an abundant supply of soil water, they often exude water from the edges of their leaves. Greenhouse plants, after a good watering, may often be in such a condition, and transpiration at such times is very low. This process of exudation from the leaves is known as guttation, and it may often be seen on

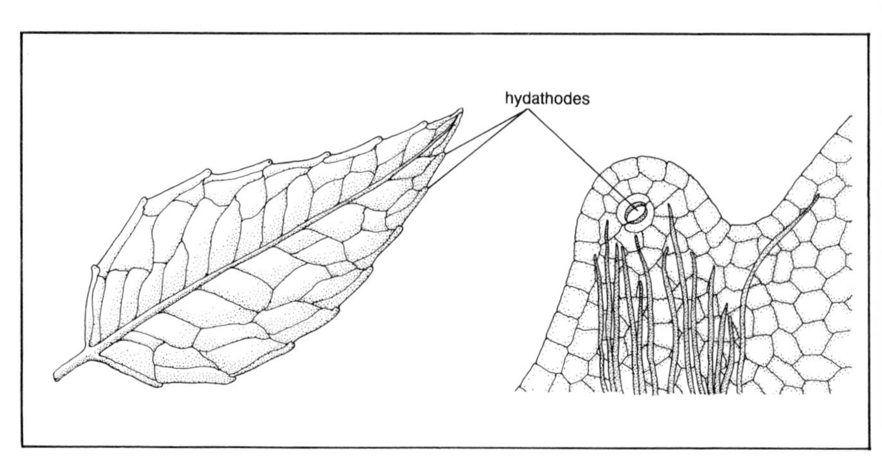

Leaf of fuchsia showing 'drip tips' or hydathodes, with enlargement of the hydathode structure (right).

the leaves of tomatoes, begonias and the swiss cheese plant (*Monstera deliciosa*). It seems probable that guttation permits the movement of water and salts through the plants even where the transpiration rate is low. It is probably linked to respiration and requires energy input from the plant for it to take place. Rates of guttation may be surprisingly high, and up to 200 millilitres of water have been recorded as being lost from a single leaf each day. Plants that habitually guttate usually have specialised water-secreting glands known as hydathodes that connect directly to the xylem vessels, though in other cases guttation occurs through the stomata.

GROWTH

All normal plants grow between the time of germination and their death, and in some cases (such as an oak tree) this growth is very considerable. Not only do they grow in length and usually girth, but they also produce numerous specialised organs and tissues such as leaves and flowers, and they also react to external stimuli such as light, or gravity, by growing towards or away from them. The system by which plants control all these forms of growth, largely by hormones, has become clearer in recent years and the incredible complexity and flexibility of growth is at last becoming known.

Primary growth

When a seedling germinates, the root and the shoot grow with remarkable rapidity pushing their way through the soil or up into the air. The transformation from seed to mature plant is one of the most extraordinary transformations in nature, but how is it actually achieved? As we have already seen, the region of greatest growth in the root or shoot is in the meristem or growth tissue just behind the apex. The actual process of growth takes place in two stages, cell division and cell elongation, and these two are followed by a third, differentiation, in which the basic cell types change into distinctly recognisable tissues (groups of similar cells) and organs, e.g. a leaf.

Cell division is the process in which one cell replicates itself to produce two, and this takes place in the zone of cell division at the top of the meristem (i.e. the region nearest the apex of the root or shoot). The process of cell division, the real basis of growth, is a complex and extraordinary process which it is worth looking at in more detail, as it is shared by all growing cells throughout the plant world. As we have seen, the cell contains within it a viscous fluid called cytoplasm which contains a single nucleus, the control centre. In actively dividing cells, the nucleus is normally very visible, and it

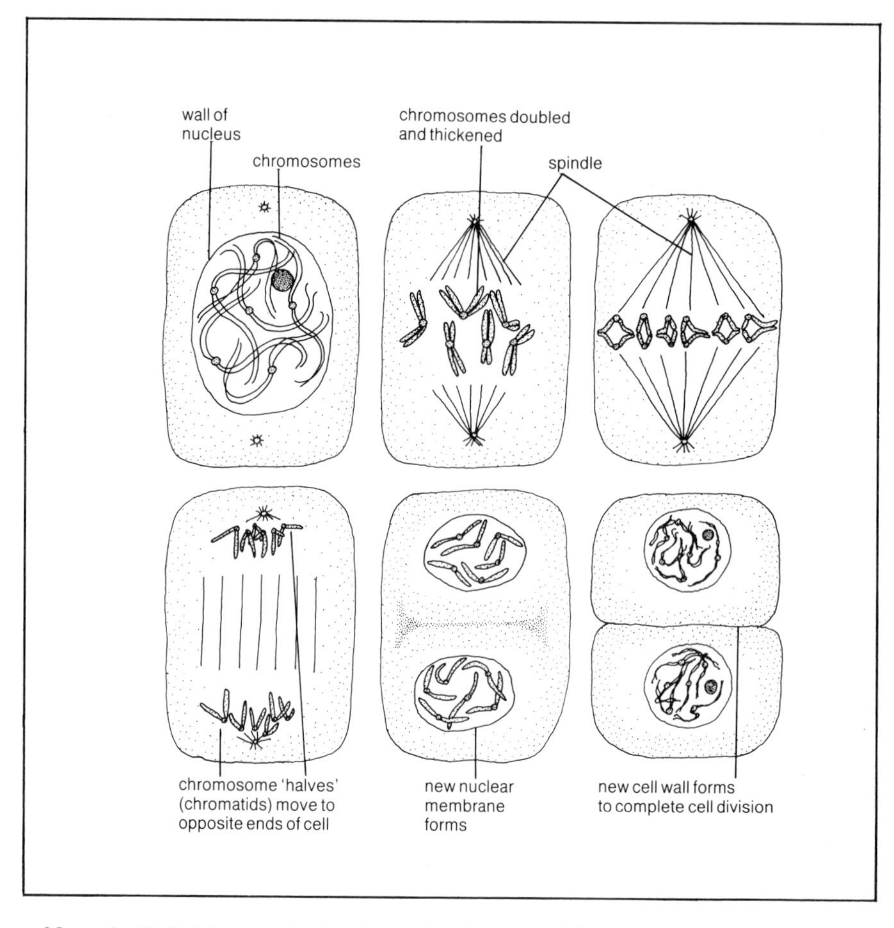

Normal cell division or mitosis: the production of two identical cells from one parent.

can be seen, as the time of cell division approaches, that within the nucleus there are a number of paired strands, the chromosomes. (There may be anything between 6 and over 100 of these chromosomes, depending on the species of plant.) Normally these are invisible within the nucleus, even under the microscope, but just before cell division the chromosomes become clearly visible as they shorten and thicken. The purpose of cell division is to produce two cells identical in every respect, and this process begins with the chromosomes, each of which split before division into two loosely associated identical halves, known as the chromatids. Then the wall of the nucleus breaks down and these paired chromatids move in an orderly

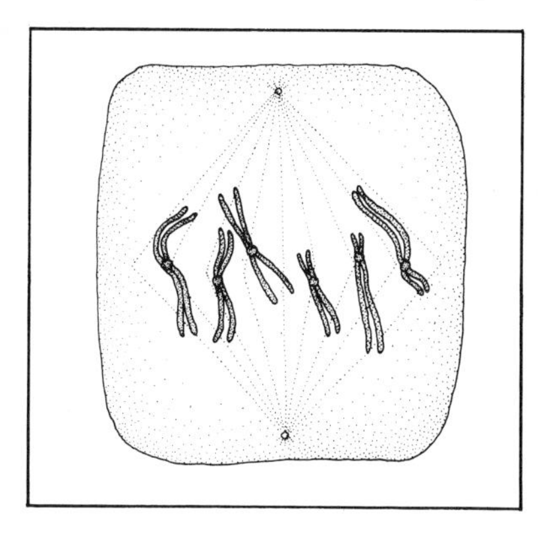

Enlargement to show structure of the divided chromosomes during cell division.

fashion to the centre of the cell where they lie across what has become known as the equator, for obvious reasons. The paired halves now split from each other, and at the same time strands of cell cytoplasm are forming a spindle stretching away to each end of the cell. Then the chromatids move along these fibres, one half of each pair to each end of the cell. Once there, a new nuclear membrane forms around them, and each one divides so that there are now two nuclei within the cell, each with a full complement of chromosomes, doubled up ready for the next division. At the same time all the other parts of the cell divide, and all the other tiny functional organelles (little organs) in the nucleus divide, or are simply split in number if there are several of them, so that each end of the cell has a full complement. Meanwhile, the 'spindle' disappears and a new cell wall forms along the middle of the cell, creating two new ones. This process of cell division for growth is known as mitosis, somewhat different from the process of cell division known as meiosis which precedes reproduction (i.e. the formation of the ovules or pollen grains, *see* Chapter 4).

In the meristems, or regions of active growth, this process of cell division is constantly taking place (using up a large amount of energy) so that the number of cells in the shoot or root is constantly increasing, and this accounts for part of the visible growth. The remainder is accounted for by cell elongation or enlargement, and this takes place in a zone behind the meristem known as the zone of elongation, where no cell division takes place but where each new cell may be enlarged up to 1,000 times its original volume. It is this which promotes the rapid lengthening of a shoot or root, and it depends mainly on the absorption of water which fills the cell vacuoles and extends the elastic cell walls to their limits.

After elongation has taken place, the cells begin to differentiate. Before this phase of differentiation they are all more or less identical, but gradually they change to take on their mature characteristics—for example, the outer layers will develop into the epidermis, some of the others will develop into vascular tissue, and others hardly change at all. The way in which cells change, and what controls which ones change, is still not fully understood, though this and other processes in the plant are beginning to be understood as the complexity of the plant hormone system is becoming better known (*see* p. 62).

In addition to this form of growth in which the cells are produced in the growing tips to cause stem or root elongation, there is another pattern of growth that takes place in woody plants to cause increase in girth, known as secondary growth or secondary thickening.

Secondary growth

It is well known that the age of felled trees can be roughly estimated by the number of growth rings, but clearly these cannot have been caused by the growth taking place at the shoot tips. Instead this form of continuing lateral growth (i.e. increase in girth) is caused by a quite different meristem known as the cambium. Immediately below the bark of a woody stem is a layer of

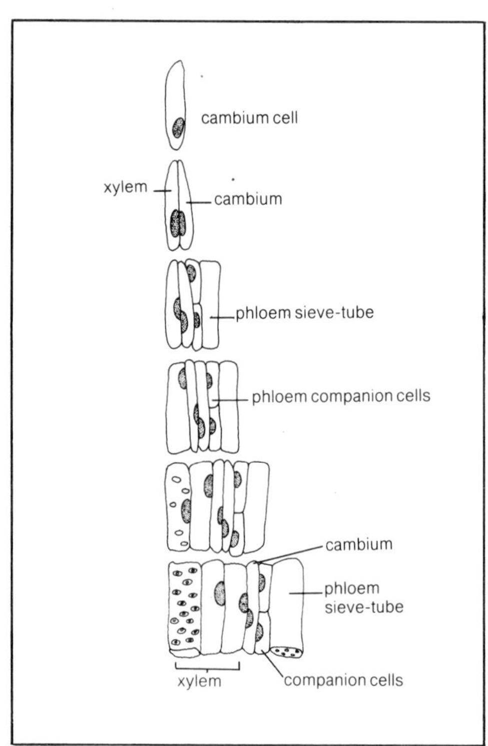

Diagram showing production of xylem and phloem by 'secondary thickening' from cambium cells.

phloem (food-conducting cells), normally very thin and often passing unnoticed. The remainder of the stem or trunk is made up of the 'wood' which consists almost entirely of thickened xylem vessels. Between the outer layer of phloem, however, and the inner core of 'wood', there is a layer of cells, one cell thick, which has been responsible for producing all the tissue of the trunk, and this layer is known as the cambium. This single layer of cells divides continuously, unless conditions are unsuitable (e.g. in the winter), to produce a layer of xylem cells on the inside and a layer of phloem cells on the outside. Thus it is always moving outwards away from the centre of the trunk. Because growth speeds up in the late spring and summer, slows down in the autumn, and ceases in winter, distinctive bands are produced each year, and it is these that can be counted, in the xylem, as annual 'rings'. At the same time, the width of the ring reflects the conditions the tree was growing under, and old trees portray the history of the wood they are growing in; fellings, competition, etc. may all be reflected in the size of the rings. Exceedingly old trees, such as bristle-cone pines in the USA, have proved useful in dating other events in that the pattern of rings in wood found in archaeological sites can be compared with still-living trees and thus precisely dated. This science is known as dendrochronology.

The control and organisation of growth

As far as we know, there is no one overall control centre in a plant, yet the small group of growing cells in the seed of, for example, a groundsel plant (*Senecio vulgaris*) produce millions of cells organised together into a coherent organism which is quite clearly recognisable as a groundsel plant, and the same applies to all other species. Those cells in the seed contain all the information, in the chromosomes, to tell them that they are a groundsel plant. What, however, determines how the plant achieves its final form, or which cells produce leaves? Why do the cells aggregate to produce recognisable shapes, and why are the flowers yellow?

The answers, unfortunately, are by no means clear, and this is one of the biggest areas of doubt in the study of plants. Some of the factors are now known, but how exactly they all work together is still uncertain. Firstly, the cells contain genetic information derived from their parents; this information is specific to one species, in that all plants of a species share certain characteristics, though at the same time it is variable within limits in the same way that human beings are all slightly different. Secondly, the plant is greatly affected by the environment in which it is living, for example by light, water, food availability and so on. Thirdly, there is a group of substances known generally as plant hormones, which singly or in combination affect and perhaps control many aspects of plant growth and development. The hormones themselves are affected both by the genetic material of the plant and by the environment, but what exactly determines how much of each hormone is produced and where it operates is still far from certain.

Three main groups of plant hormones are now known: the auxins, the gibberellins and the cytokinins. The auxins are generally active in the stimulation of longitudinal growth (i.e. increase in length), both of stems and roots, but they are also involved in fruit growth, determination of sex, initiation of root formation, leaf fall and many other events or processes. One of the best known of the activities of auxin is the suppression of lateral buds by a growing tip—it is well known that removal of an apical bud allows the lateral buds below it to develop (gardeners use the technique to promote 'bushiness') and it is now known that auxins produced by the apical bud are moved down the plant and prevent the lateral buds from developing. Similarly, it is well-known to gardeners that a cutting roots best when a leaf or bud is present on the stem, and it seems that auxins produced by the bud or leaf are translocated down the stem to the base, where they collect as they can go no further, and there they promote root initiation. Some hormone rooting powders are based on auxins and work in a similar way.

Gibberellins seem to be involved in cell division, growth by cell elongation, leaf growth, flowering, seed dormancy, inhibition of root initiation, inhibition of bud growth (in cooperation with auxins), and so on. Their most striking effects are on stem elongation and they can increase growth by four or five times, though they are clearly involved in numerous other processes, sometimes working with and sometimes against other hormones.

The third group of hormones, the cytokinins, are mainly active in promoting cell division in stems, roots, leaves and fruits. They may work with the other hormones or, at times, against them as in the case of bud dormancy which auxins encourage while cytokinins break it.

In addition to this system of hormones, there are also chemicals in the plant whose main function seems to be to prevent or reduce the activity of hormones where they are not required. So, the plant is a complex system of interacting chemicals which work together in some situations, and in opposition in others, but the whole works in perfect harmony to create a unified, functional organism.

PLANT BEHAVIOUR

In addition to all the internal controls that keep the plant growing in the way that it should, plants respond in various ways to outside stimuli, and

*Many flowers react to low light levels by closing their flowers, such as this group of wood anemones (*Anemone nemorosa*) which have closed up for the night.*

although it rarely takes place at any great speed, this is best described as plant behaviour. Plants may respond in particular ways to light, gravity, water, temperature and touch. The best-known example of these involves light.

If a box of seedlings is placed near a window or other light source, they will be seen to have bent towards this light source within a few days. If the box is then turned around, the seedlings bend back, towards the light again. So, something about the light makes the seedlings grow towards it, and it has been shown that this is primarily due to the effects of auxin. It seems that auxin is moved to the side of the stem away from the light, where it causes the cells on the darker side to elongate more, and thus the plant grows towards the light. Most stems and growing shoots react in this way. A directional growth response to a stimulus like light is known as a tropism, and shoots are therefore positively phototropic as they grow towards the light source. Some roots, in contrast, are negatively phototropic and grow away from the light. A few plants seem to exhibit negative phototropism in their above-ground parts, for instance, the ripening fruits of ivy-leaved toadflax (*Cymbalaria muralis*) grow away from the light, allowing the seeds a better chance of being dropped into a crevice in a wall, which is the main habitat of this plant. One other interesting behavioural response to light is shown by the so-called compass plants, for example the North American compass plant (*Silphium laciniatum*). These plants continuously align their leaves at a varying angle to avoid presenting leaf surfaces to the sun at the hottest part of the day. Light also affects almost every other aspect of plant growth, including its speed and form of growth, its flowering and so on, but these are not, strictly speaking, behaviour and are not considered further here.

Perhaps surprisingly, gravity affects the way plants grow. Shoots grow upwards and roots grow downwards, yet if you invert a plant it rapidly alters its growth pattern to allow the roots to continue to grow downwards, and the shoots to continue to grow upwards. This growth response to gravity is known as geotropism; shoots grow away from a gravitational pull, and most roots grow towards one, so roots are said to be positively geotropic while shoots are said to be negatively geotropic. The advantages of this to the plant are obvious, especially in a germinating seed away from any light source. It seems that, once again, the auxin plant hormones are involved in promoting this reaction, though it is not quite clear how.

Gravity also affects the way that plants develop. Experiments with plants grown in the absence of gravity show that they will develop quite differently in such conditions, but more relevant is the use made by gardeners of the effects of gravity; if a shoot of, for example, cherry or apple is trained or held horizontally the effect, caused by gravity, is that shorter internodes and more flowers and fruit result. Espalier fruit trees, trained along walls or fences, thus not only make good use of space but also produce more fruit. It appears that there is some redistribution of auxin to the lower side of the stem, which possibly allows the lateral buds to develop, but it is doubtful if this is the full story.

Two other growth movements that some plants show are the growth of some roots towards a source of water (positive hydrotropism) and the curling of the tendrils of some climbers around any solid object with which they come into contact (positive haptotropism).

Some rather different behavioural responses which do not involve growth but are reversible non-directional responses caused by a stimulus such as light are known as nastic movements. Good examples include the opening and closing of flowers in response to sunshine (e.g. dandelions (*Taraxacum* spp.) which open only during the day, or goat's beard (*Tragopogon pratensis*) which opens until midday), and the response of the sensitive plant to being touched. The sensitive plant (*Mimosa pudica*) responds very rapidly to being touched by collapsing its leaves and leaflets completely. These recover after about 15 minutes to their original position. The exact mechanism of this is not known, though it is certainly associated with loss of water pressure in groups of cells at the base of the stems. A similarly rapid reversible response can be seen in the reaction of the leaves of the Venus flytrap (*Dionaea muscipula*) to the touch of a fly. As soon as a trigger is touched, the leaves snap shut and catch the fly.

So, plants are not quite the inert objects that they may sometimes seem. As we shall see later (Chapter 6), they can even detect day length to a matter of minutes, and their internal mechanisms are infinitely more complex than one might first imagine.

4 · REPRODUCTION

To survive and do well, flowering plants have to reproduce themselves successfully. In addition, as we have seen, it is beneficial to the species if reproduction can be carried out by sexual means (i.e. the male parts of one plant combine with the female parts of another) because this introduces greater variability into the resulting offspring, which in turn allows more opportunity for the species to evolve with its environment. To achieve this, male and female parts have to be produced, and the cells of the male parts of one plant have to be transferred to the female cells of another plant (this process is known as pollination). Flowers are produced in order to achieve this aim.

At the same time, it may often benefit the plant if it can spread without the rigmarole of producing flowers and attracting male cells from other plants, for example when heavily grazed or mown, or in very severe climatic conditions, and many plants have evolved alternative methods of reproduction. These are known generally as vegetative or asexual reproduction, and plants may combine both methods or rely mainly on one or the other.

SEXUAL REPRODUCTION

Flower production, pollination, and fertilisation, leading to the development of the seeds and fruit, are the main stages in the process of sexual reproduction.

Flowers

Everyone is familiar with flowers. We use them for decoration and perfume, and see them every day in the garden or in the countryside. They are so

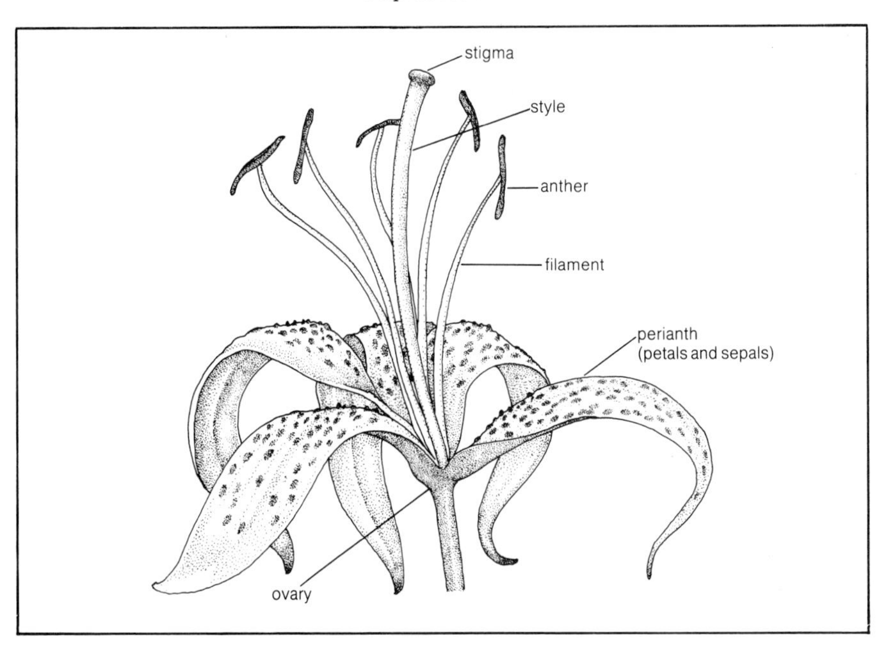

A typical flower showing the main reproductive parts.

familiar, in fact, that it is easy to forget that they have a purpose from the point of view of the plant, and they do not exist solely to give pleasure to humans! Their variety is astonishing, from the complex structure of orchid flowers, through the composite flowers of the daisy family, to simple flowers that are hardly distinguishable from leaves. Yet they all serve the same purpose: to allow the plant that bears them to reproduce sexually and produce seeds. Much of the variety and splendour of form has evolved because most flowering plants rely on insects to transfer the male cells (pollen) to the female parts of another plant, and, to this end, the insects must be attracted to visit as many flowers as possible. So the variety of form has evolved to attract a range of different insects, and to outdo possible competing distractions.

Before looking in more detail at the ways in which flowers achieve their purpose in life, it may be useful first to understand what comprises a flower. Most flowers have a basic structure that is repeated throughout the flower world, though some specialised types have dispensed with, or combined, one or more of the basic parts. A simple, characteristic type of flower, such as that of a wild rose, consists of an outer ring of protective leaf-like structures that enclose the bud before it opens and prevent frost or other damage.

These are known as the sepals (or, collectively, as the calyx) and though they are normally green they may occasionally be the same colour as the petals. Sometimes they drop off shortly after the flower has opened, e.g. in poppies, though often they persist right through to fruit formation, e.g. the green 'stars' attached to tomatoes or strawberries (both of which are fruits) are the persistent sepals. Generally, though, they play little further part in reproduction once their role as protectors of the developing flower has been completed.

Immediately within the sepals lies another whorl of leaf-like structures, often highly modified and coloured, known as the petals (or, collectively, the corolla). These represent the part of the flower that is most familiar, and is, indeed, generally thought of as 'the flower'. In fact, their role is almost entirely that of attracting insects and helping them to carry out pollination, and they do not form part of the sexual apparatus—the real flower—at all. In flowers that rely on wind or water to transfer their pollen, the petals may be almost or completely absent, e.g. in shoreweed (*Littorella uniflora*). Normally there are the same number of petals as sepals, usually between three and six, though this is not always the case. In many flowers, parts of the petals have become modified to form nectaries. These are organs or groups of cells which produce a sugary substance (nectar) attractive to insects (and the basis for the sweetness of honey). The nectaries may be simple sac-like structures at the base of the petals, as in the flowers of buttercup, or they may be much more complex structures like the long spurs of columbine (*Aquilegia vulgaris*) or many orchids. Often the nectar is so far down a spur or other structure that it is only accessible to long-tongued

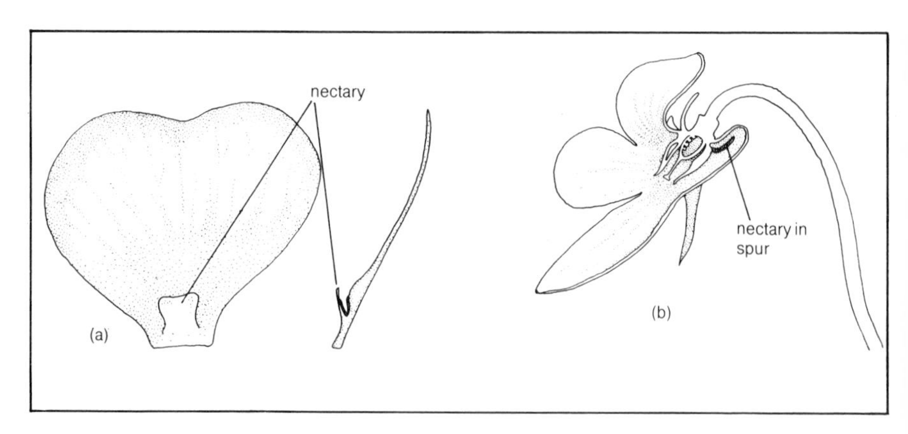

Nectaries of (a) buttercup and (b) violet. Different insects can make use of each type.

insects (e.g. hawkmoths), thus reducing the number of insects that will visit the plant and increasing the likelihood of cross-pollination. (If the nectar is available to any insect, the chances are that the insects will visit flowers of a different species next because they are not specialised feeders.)

The petals and sepals together are sometimes known as the perianth, especially where they are very similar in form and colour, as, for example, in the bluebell flower. Generally, the 'dicots' (*see* Chapter 1) have petals and sepals in fours or fives (or occasionally multiples of these), whereas the 'monocots' have them in threes or multiples of three.

Enclosed by the sepals and petals, and protected by them, are the working parts of the flower, the male and female organs. Most flowers contain both male and female parts, though in some species they are in separate flowers on the same plant (e.g. most sedges) and sometimes there are separate male and female plants (e.g. holly in which only the female bush bears the berries, or the red campion, *Silene dioica*). Such plants are called dioecious.

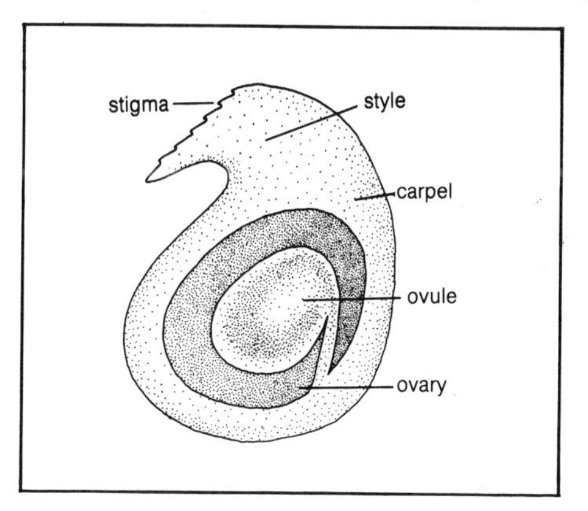

Cross-section of a generalised carpel with one ovule (e.g. buttercup).

The female part of the flower usually forms the central portion and it consists of one or more carpels, each of which contains one or more eggs, or ovules, surmounted by a style and stigma. The stigma is the receptive surface on which pollen grains (the male parts) can land and grow, while the style is simply its stalk. Flowers may have numerous single carpels, like a buttercup, or one or more may have joined together, sometimes with a common style or sometimes with separate styles. The ovary may contain one ovule, like the plum or cherry with its single 'stone' (the single seed, developed from one ovule) or it may contain several ovules like a pea pod, or even thousands, as in an orchid ovary. Plants have evolved different strategies for getting their seeds dispersed, and the number of ovules and the amount of food dispersed with them are a reflection of these strategies.

The male parts are usually found in a ring around the central female parts, and they consist of pollen-bearing structures called stamens. Usually these are formed from a stalk called a filament, and the actual pollen-bearing organ called the anther. The anther is filled with a yellow dust-like powder known as pollen, consisting of thousands or millions of tiny grains, each containing a male cell. The actual structure of the pollen grains, when seen under the microscope, is enormously variable, yet characteristic for each species or genus of plant. They may be globular, star-shaped, spiny, pitted, smooth, and so on, yet each is recognisable as belonging to a particular species or group of species. This fact is the main basis of the science of palynology: pollen grains which landed on the surface of bogs many thousands of years ago have been preserved and incorporated into the peat as it formed. Because the peat builds up in layers, the pollen preserved at each level can be identified and dated, and the surrounding vegetation at the time can be guessed at. Much of our knowledge of the composition of the woodland covering Northern Europe after the last glaciation, and its subsequent clearance by man, has come from the study of preserved pollen grains.

The male and female parts (the stamens and the carpels) are the only parts of the flower that always occur (though not necessarily together, as we have seen) and the petals, sepals, nectaries and so on are additions that may be present for specific purposes, but are not necessarily so.

Flowers may be produced singly, like tulip flowers, or they may frequently be aggregated to form recognisable groups known as inflorescences. The purpose of these aggregations normally seems to be to aid in attracting potential pollinators, especially insects, and the inflorescence may have specialised parts within it, such as bracts (coloured leaves) which help to create the effect. The red 'flowers' of the familiar house plants, poinsettias, are actually coloured leaves below the true flowers, which are tiny. Often, though, the inflorescence must serve some other purpose, for the catkins of trees and some bushes are inflorescences, yet their pollination is carried out entirely by wind. In this case, the aggregation of numerous flowers probably allows a more flexible form to be developed allowing better distribution of pollen.

---

*The beautiful catkins of hazel (*Corylus avellana*) are simply the male flowers; the female flowers are tiny red rosettes, visible on the left of the picture, which receive the pollen and eventually grow into a hazel nut.*

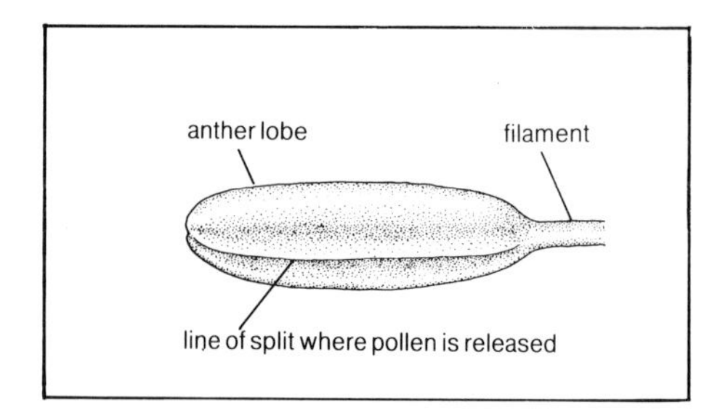

Greatly enlarged anther lobe.

Inflorescences may take many forms. The simplest involve a series of flowers equally placed up a main stem: a spike in its basic form, where the flowers are unstalked, or a raceme where the flowers are stalked. When the stalks of the flowers on the raceme become shorter and shorter towards the top, so that all the flowers end up at the same level, it is known as a corymb. Where the inflorescence has many branches, with each branch ending in a flower and growth continuing from laterals produced below, it is known as a cyme.

A more familiar type of inflorescence is the umbel, where all the flower

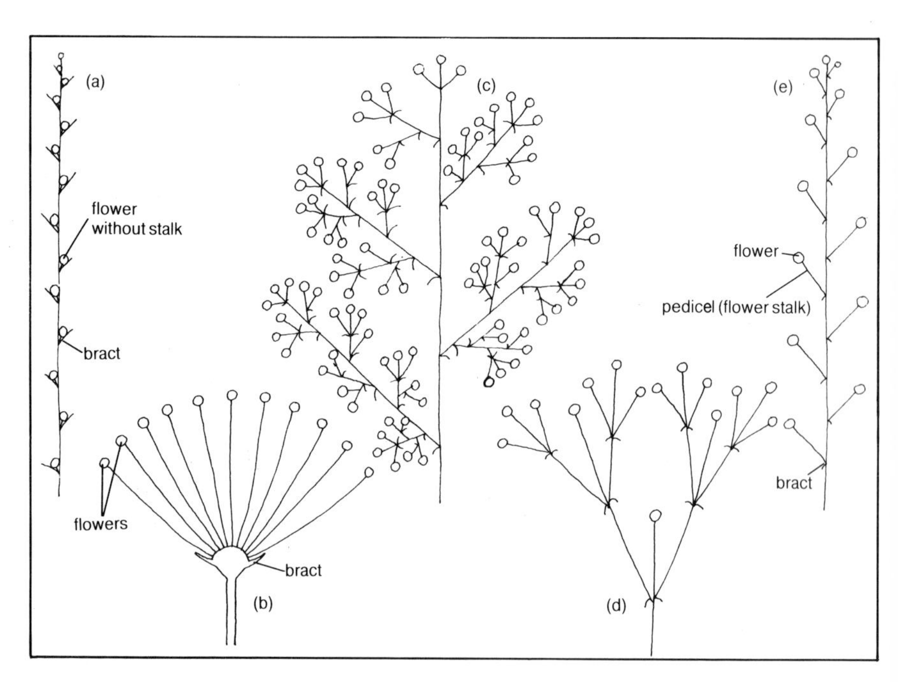

Types of inflorescence: (a) spike; (b) umbel; (c) panicle; (d) cyme; (e) raceme.

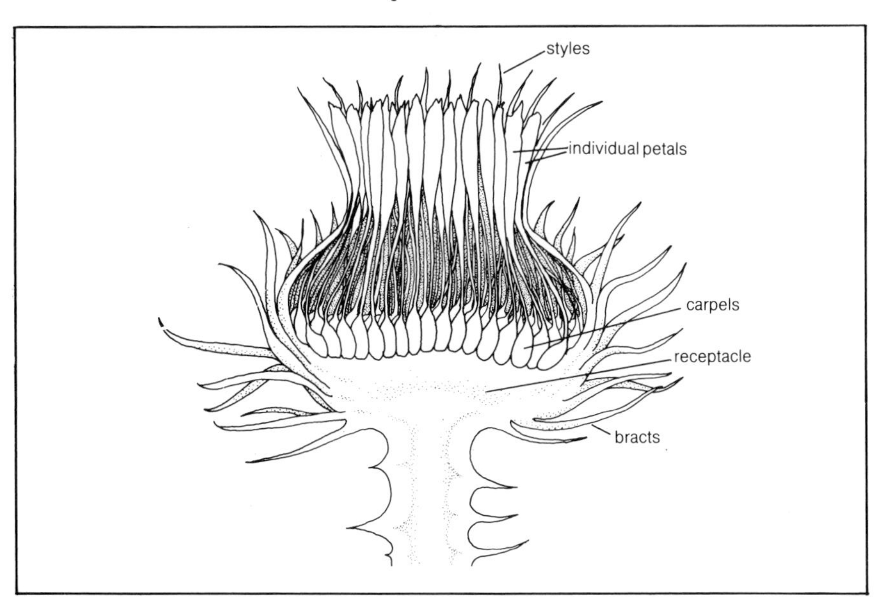

Section of a composite flower (one of the thistles) to show its complex structure.

stalks arise from a single point at the end of the stem to create an effect like the spokes of an umbrella. Virtually all the inflorescences of the carrot family, the umbellifers, such as hogweed or angelica, have this form of inflorescence, hence the name of the family. Within such inflorescences, there may be specialisation amongst the flowers so that, for example, the outer flowers are sterile but very large and attractive, whereas the inner flowers are smaller but fertile. The inflorescence acts as a coordinated unit, with a greater chance of attracting insects.

One other form of inflorescence is familiar to most people, though not usually recognised as such. The 'flowers' of the dandelion family (the Compositae) are actually complex inflorescences, often containing several different types of flower. The 'flower' of a daisy (*Bellis perennis*) for example consists of sterile white ray florets around the outside (to provide the visual attraction) and yellow fertile disc florets in the centre, with much reduced petals.

Whatever their structure, the feature that is most striking about flowers is usually their colour. This colour is produced by a variety of pigments occurring singly or in combination, either in special bodies within the cell known as plastids (like the chloroplasts in the leaf which contain the green pigment, chlorophyll), or distributed throughout the cell. The red and yellow pigments, which can account for most shades of yellow, orange and red, usually occur in plastids, while the commonest of the dissolved pigments are the anthocyanins. These account for the blue, purple, magenta, brown and even black in flowers. The exact shade of colour depends on the concentration and combination of pigments and the acidity of the cell-sap—anthocyanins are red in acidic conditions and blue in alkaline conditions, and some flowers change colour according to the acidity of the soil they are growing in (for example, hydrangeas). White flowers are not coloured by a pigment at all. The whiteness is produced by numerous tiny air spaces between the cells, in the same way that froth appears white. Many flowers that are normally coloured, often highly coloured, produce albino forms which are white, or nearly so, simply because of a lack of pigment caused by a genetic mutation. It is noticeable that many such albinos retain a shade of yellow or green if either of those pigments is present, and it is usually the reds and blues that are lost.

Pollination

The basic function of flowers, however, is to achieve sexual reproduction by transferring the male cells (pollen) from one plant of the species to the female parts (stigma) of another. The process by which this is achieved is known as pollination. To achieve the maximum possible variety in the offspring (and therefore greater capacity for evolution) cross-pollination is preferable, i.e. the pollen landing on the stigma and germinating will be from a different plant of the same species rather than from the same plant. Most plants are habitually cross-pollinated, and there are many ways in which this is achieved, though some plants have dispensed with it altogether and habitually pollinate themselves, sometimes before even opening. There are really two different strategies for survival: plants that are cross-

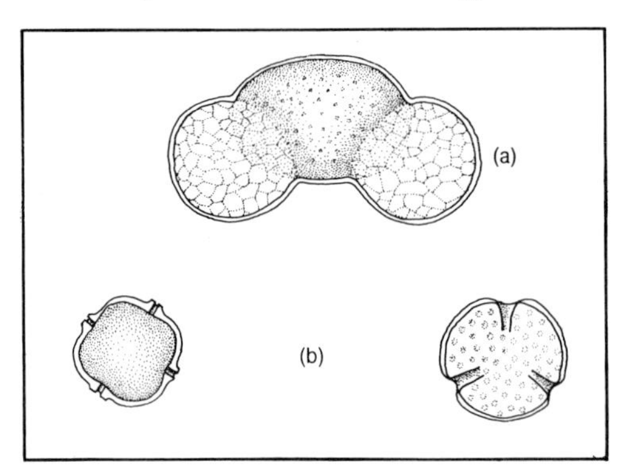

Pollen grains, greatly enlarged: (a) wind-pollinated type; (b) insect-pollinated type.

*Members of the mallow family, such as this shrubby althaea (*Hibiscus syriacus*) prevent self-pollination because the stigmas do not open until the 'tube' of yellow anthers has withered.*

pollinated run the risk of low seed-set (i.e. few of the ovules are fertilised) through lack of pollinators, lack of nearby plants of similar species, poor weather etc., but because the offspring are more varied they have a greater chance of survival; plants that are self-pollinated have a high percentage of seed-set, but there is less variation amongst the offspring, i.e. the former has long-term benefits, and the latter has short-term benefits.

There are three main ways in which cross-pollination normally takes place: by insects (and other invertebrates, or even vertebrates occasionally), by wind, and by water.

INSECT POLLINATION

Insect pollination is by far the commonest method, and accounts for some 80 per cent of flowering plants. Because the flowers and the insects have become mutually dependent—the flowers needing pollination and the insects needing food—they have evolved together to some extent, and there is an enormous variety of flower types that have evolved to attract different insects. Pollination is a very wasteful process in that enormous numbers of pollen grains are produced (usually well into the millions) yet only a very few reach their target and germinate. Insect pollination reduces this uncertainty to an extent, especially if an insect that only visits one species of plant can be attracted. The method that a plant has evolved depends largely on the ecology and form of the plant. Alpines and arctic plants, for example, tend to be insect-pollinated or self-pollinated, for, although there is invariably plenty of wind, there is a very low density of plants. The chances of pollen getting from one to the other by wind alone are therefore very low. If insects can be attracted, however, they will seek out the favoured plants and move from one to the other. Trees, in contrast, tend to grow in groups (i.e. woods) and they have the advantage of height, so in most cases they have adopted wind pollination.

Clearly the most likely effect of an insect visiting a flower or inflorescence is that pollen will be transferred to the female parts of a flower on the same plant or even to the stigmas of the same flower. This would defeat the purpose of sexual reproduction, and is effectively the same as self-pollination, so various methods have evolved to prevent this from happening.

Firstly, the male and female parts may be borne on completely separate plants, as in holly or some willows. This, obviously, ensures that they are cross-pollinated, though difficulties may arise if there are considerable distances between the plants.

Secondly, the plant may ensure that within one flower, and as far as possible within an inflorescence, the male and female parts mature at different times. In the flowers of mallow (*Malva sylvestris*), for example, the anthers form a tube around the female stigma, completely enclosing it. When the anthers have shed their pollen, and withered, the stigma begins to grow and pushes its way through the remains of the stamens and opens to receive pollen from other flowers. In this instance, where the male parts mature first, the process is known as protandry. In other plants, like the plantains (*Plantago* spp.), the reverse is the case with the female parts maturing first. This is known as protogyny. Primroses have evolved a slightly different method involving two types of flower, the famous pin-eyed and thrum-eyed flowers, which occur in roughly equal proportions. Thrum-eyed flowers have a ring of stamens at the mouth of the petal tube, and a short stigma below, whereas pin-eyed flowers have a much longer-styled stigma reaching the top of the tube, with a ring of shorter stamens below this. The flowers of one plant are all of one or other type, but not both, and they can only pollinate flowers of the opposite type.

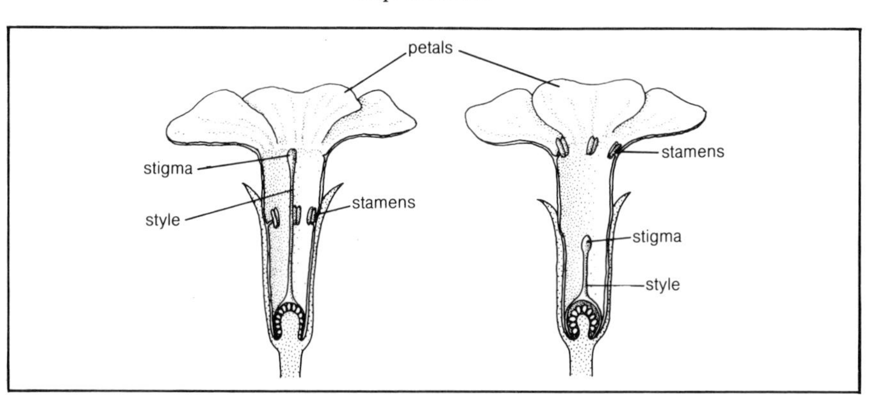

Types of primrose flower: pin-eyed (left) and thrum-eyed.

Thirdly, some plants have developed a self-sterility or incompatibility system. This is genetically controlled, and the result is that pollen cannot germinate on a stigma of the same genetic constitution, thus preventing self-pollination. In some cases, this may be even more subtle, and the stigma becomes receptive and allows germination of its own pollen if no other suitable pollen has arrived, as a last resort. In theory, at least, this combines the best of both worlds.

Plants attract insects by colours (which are not necessarily visible to the human eye), scents (which are not necessarily pleasant), and food such as nectar or pollen. Nectar is the primary attraction, and colour and scent are simply ways of indicating its presence to the insect, and attracting it from a distance. Nectar is an excellent energy source for insects, and, for many, one of their main foods. Some plants, like poppies, secrete no nectar at all, but rely on insects that come to eat the pollen, inevitably carrying some of it away with them.

Insects perceive colours in a different way to humans, so it is not always easy to understand why some flowers are more attractive than others. In general, insects can distinguish colours but they are much more sensitive to ultra-violet light than we are, and less sensitive to red. Blues, mauves, purples and yellows are the most popular colours for flower-visiting insects. To succeed in attracting insects, flowers have to stand out from the

---

Overleaf: Many insects visit flowers to eat the pollen rather than to drink nectar. This longhorn beetle (Strangalia maculata*) will effect pollination by inadvertently transferring pollen to the next flower that it visits.*

surrounding foliage and, because of the insects' different vision, they may do so in a way that is not apparent to humans.

Flowers that open in the evening or at night, and are pollinated by nocturnal insects, e.g. the night-flowering catchfly (*Silene noctiflora*) tend to be white or very pale in colour to be more noticeable and reflect as much of the available light as possible. Some flowers have lines or patterns on the petal (not necessarily visible to us) which guide the insect to the nectar source, and it has been shown experimentally that insects will follow these patterns. They are generally known as 'honey guides'.

Scent, too, plays a major part in attracting pollinators. In many cases, it simply reinforces the visual aspect and adds to the attraction. In other cases, it has been shown that an insect may be attracted to a flower visually but will not enter unless the scent is right. Many night-flowering or evening flowers are strongly scented to help insects to locate them in the dark. Moths, in particular, are very sensitive to smell and can detect an attractive scent from a considerable distance, an essential quality, if you consider the difficulties of finding both a mate and a source of food in the dark without it!

Although we tend to think of flowers as being sweetly scented, there are also a group of flowers that produce what is, to us, an offensive odour. Such plants tend to attract carrion feeders, and especially flies, rather than moths or butterflies. Often they are coloured dark red or purple to add to the impression of rotting carrion, but in contrast to the nectar-producing scented flowers they are delusions, providing nothing for the insect. A good example is the familiar lords-and-ladies, or jack-in-the-pulpit (*Arum maculatum*), which gives off an offensive odour. The flies which are attracted to the flower push down through the ring of stiff hairs in a tube to get to the source of the scent, the flowers which are protected inside a coiled bract known as the spathe. Once inside the chamber, they are trapped and their attempts to free themselves result in the transfer of pollen (preferably from a previous plant) to the stigmas. When pollination is sufficiently complete, the ring of hairs withers and releases those insects that are still alive to go and visit another flower. There is no apparent advantage to the insect, and the mechanism depends on the availability of large numbers of small carrion-feeding insects at the right time of year.

These are the basic ways in which flowers attract insects as possible

---

*Previous page: The attractive white flowers of the white deadnettle (*Lamium album*) are adapted to larger insects such as bumble bees or to night-flying insects attracted by the white colour.*

pollinators; the whole beautiful and complex world of flowers, with the exception of cultivars, has evolved for the attraction of insects—our appreciation of it is an incidental feature! With so many species of plants in existence, however, how can one plant ensure that an insect returns to a plant of the same species while the pollen is still fresh enough to effect pollination? Many ways have been adopted, varying from simply producing lots of nectar in a showy flower to a highly complex relationship with one species of insect.

The 'generalist' strategy, typified by the buttercup (*Ranunculus* spp.), relies on a good supply of nectar available to almost any nectar-eating insect and a bright yellow, highly attractive flower. The chances of successful cross-pollination are not as low as might be imagined, and the successful seed-set of the buttercup bears this out: firstly, buttercups tend to have a short, defined flowering season, quite early in the year, and many possible competitors are not in flower at the same time; secondly, different insects or groups of insects have preferences for particular colours, and those that like yellow will constantly return to buttercup flowers; thirdly, the nectar of many flowers will be unavailable to the general feeders, thus forcing them to concentrate on one or two species that do have available nectar. So, in a given meadow at a particular time of year, the range of competing flowers may actually be very small, making successful cross-pollination very likely.

This is the basic strategy of attracting insects. Much more interesting, though, are some of the remarkable adaptations that have evolved to attract a restricted range of insects, or even one species only. In such situations, the plant and the insect become totally dependent on each other, resulting in highly accurate cross-pollination, though there may be problems where external factors affect the populations of one of the two evolutionary 'partners'. The next step up the evolutionary ladder involves the development of structures that ensure that pollen is deposited on a particular place on an insect which will come into contact with the stigma of a flower if the female parts are sexually mature. Generally, these types—particularly common in the mint family (Labiatae) and the pea family (Papilionaceae)—are attractive to groups of insects, especially bees and hoverflies. The sage (*Salvia verbenacea*) for example, has a large specially-shaped lower petal which acts as a landing stage for insects of a particular size and shape; there are two stamens which ripen before the stigma, and these have a special sterile lobe below the fertile part of the anther. The immature style is high above the insect at this stage, well out of the way. When the insect lands on the petal and forces its head into the corolla to get at the nectar, its body pushes against the lobe which acts as a lever and brings the fertile pollen sac down onto its back, covering it with pollen. When the bee, or other insect, visits another flower which is more mature, the style and stigma (which lengthens as it ages) will be in exactly the same position as the anther was in the less mature flower. By this means, the pollen is accurately transferred from one flower to another.

The best known and most impressive of the highly-evolved insect-pollinated plants are the orchids. Many of the temperate orchids, including several that occur in Britain such as the spotted orchids and marsh orchids (*Dactylorhiza* spp.), have evolved a special mechanism that involves the insect removing the whole stamen, which then moves forward into the correct position for cross-pollination and is carried to the next plant. The flowers are complex and unusual, and they involve an enlarged front lobe for the insect to land on, a very long spur containing nectar, and specially adapted sexual parts: there is only one stamen, split into two lobes called pollinia, each of which has a sticky pad at the base. Instead of being contained in a sac, the pollen is open to the air but the grains are lightly held together with sticky threads. The stigma is a broad sticky area lying below the bases of the pollinia, but separated from them by a bump, called the rostellum. When the insect visits the flower and attempts to reach the nectar in the spur, its head pushes against the sticky bases of the pollinia. These then adhere to the insect's head, rather like two horns, standing upright. In this position, they would not touch the stigmatic surface of another flower, so, after about a minute, they bend through 90 degrees and end up pointing forward, so that they then touch the stigma of the next orchid flower that the insect visits. The delay of a minute or so is remarkable, in that it greatly reduces the chances of self-pollination. The pollinia can go on pollinating all the flowers that the insect subsequently visits, as just a few pollen grains detach on each visit. The pollinia are so sticky that it is not uncommon to see a fly or a bee in an orchid meadow with several on its forehead, buzzing about like a spaceship with antennae! Eventually they will remove them, or knock them off.

The twayblade orchid (*Listera ovata*) has a similar mechanism with a few significant differences: in this orchid, the lower petal is shaped rather like a

---

*The pink double pollen sacs of the spotted orchid (*Dactylorchis fuchsii*), visible in the photograph over the centre of each flower, attach themselves to visiting insects by means of sticky pads, and then change position so that they hit the stigma of the next flower visited.*

man, with two lobes like legs, and a central groove in which nectar is produced. A visiting insect is attracted by this nectar and follows the groove up the petal, eventually reaching the top of the petal, where the pollinia lie above the stigma, separated from it by a little projection, the rostellum. As soon as the insect touches the rostellum, however gently, a miniature explosion takes place and a tiny drop of liquid is exuded onto the base of the pollinia. Within a few seconds this sets hard and attaches the pollinia to the insect's head, and at the same time serves to startle the insect sufficiently (depending on the species) into leaving the first flower and seeking another. On doing so, it may either collect another set of pollinia, or if the flower is more mature and the rostellum has curved away upwards it will successfully pollinate the flower and many more afterwards until the pollen has been used up or the pollinia removed.

One final example of a related but even more subtle method is found in the pyramidal orchid (*Anacamptis pyramidalis*), a widespread plant of chalk grassland in Europe, including Britain. The pink lightly-scented flowers of this species have a long slender spur about half an inch (13 millimetres) long and therefore accessible only to long-tongued insects such as moths and butterflies. The pollinia are joined to a single sticky pad and the stigma is split into two surfaces on either side of the entrance to the spur. As the visiting insect alights and pushes its proboscis into the spur cavity, it is guided by two converging walls so that it passes below the base of the pollinia. The sticky base of the pollinia is in the form of an oval with its long

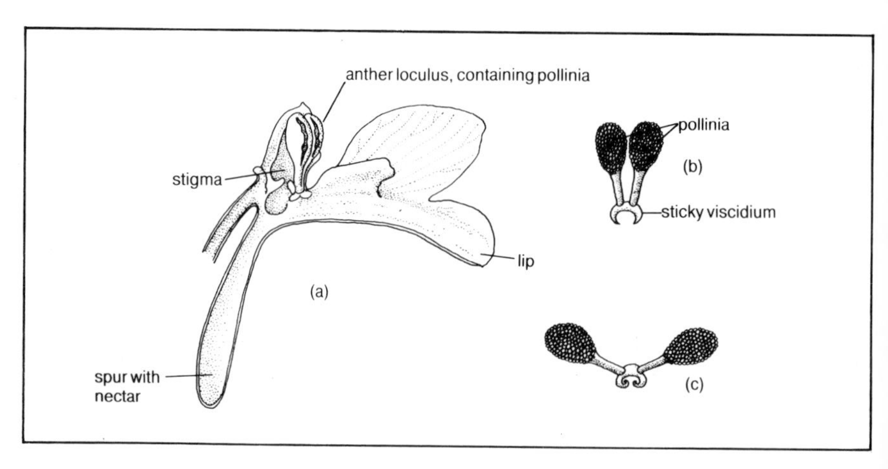

*Pollination mechanism of pyramidal orchid (*Anacamptis pyramidalis*): (a) partial section of flower; (b) pollinia removed, as if on insect head; (c) pollinia after outward movement has occurred, helping pollination of next flower visited.*

axis lying across the path of the proboscis. As soon as the sticky pad touches the proboscis, it adheres to it, and the two free ends of the pad immediately curl round the proboscis, gripping it tightly. The 'glue' then sets hard, and by the same action the two parts of the pollinia are forced to separate so that they now point outwards. So, when the insect flies off, it takes the pollinia with it, firmly attached to its proboscis but now splayed in such a way that they will touch the two stigmas of the next pyramidal orchid flower that the insect visits.

In Britain only the 'tip of the iceberg' is seen in terms of the development of bizarre orchid flowers and pollination mechanisms. In the tropics, where both the number of orchids and the variety of insects are far greater, their development and co-evolution reaches its zenith. One example will serve to illustrate this, though the variety is enormous. A madagascan orchid (*Angraecum sesquipedale*) has a spur up to 12 inches (30 centimetres) long, and a strong scent. Charles Darwin, in his study of orchids and pollination, predicted that there must be an insect with a 12-inch (30-centimetre) proboscis, long enough to reach the nectar and pollinate the flower. Some 40 years later, just such a moth was discovered and it did indeed pollinate the orchid!

Probably the most intriguing of all the bizarre attractions provided by flowers for insects is the development of flowers that so closely resemble the insects they are designed to attract that the insect is fooled into believing the flower is a potential mate. This method is widespread amongst one particular group of orchids in Europe, the genus *Ophrys*, which includes the bee orchid, fly orchid, wasp orchid and others. The fly orchid (*Ophrys insectifera*) has a flower which closely resembles the females of a wasp, *Gorytes mystaceus*. The males of the wasp emerge before the females, by which time the flowers of the fly orchid have opened. Not only do the flowers look like the requisite insect, but they also smell like them, and the male wasp is induced to attempt to mate with what he believes is a female on a flower. Frustrated, he flies off with the pollinia and visits another flower, thus effecting pollination. It is obvious, though, that this method is so specialised that it may break down under adverse conditions. Not only does the single species of insect (and it

---

*Overleaf: Fly orchids (*Ophrys insectifera*) attract the males of a small wasp by mimicking the appearance and even the scent of the female wasp. Pollination is effected as the male flies from flower to flower.*

Above: In early spring, before the leaves are out, most trees produce clouds of pollen which disperse on the wind to the flowers of adjacent trees. The picture shows pollen dispersing from the flowers of wych elm (Ulmus glabra *).*

Below: The flowers of the false oat grass (Arrenatherum elatium *) are superbly adapted to wind pollination with their long dangling anthers and feathery receptive stigmas.*

often is specialised to just one species) have to remain sufficiently widespread, but the synchronisation of life cycles has to be just right, as a real female proves more attractive every time! In Britain, which is at the edge of the range of both the orchids and the insects, the system does indeed break down, and the pollination success of the fly orchid by its sole known pollinator is only about five per cent. Some similar orchids, such as the bee orchid (*Ophrys apifera*), have developed a 'fail-safe' mechanism in which, if no suitable insect has arrived, the pollinia dangle out of their pouches and curl backwards eventually to touch their own stigma, ensuring that seed is set even if only by self-pollination.

WIND POLLINATION

Most of the remaining 20 per cent or so of flowering plants are pollinated by the wind. This strategy involves producing enormous quantities of pollen in the hope that some will reach the stigma of another plant of the same species. The pollen grains of wind-pollinated species are incredibly small and light, with a simpler structure than the pollen from insect-pollinated flowers, and are produced in enormous quantities: 210 million grains of pollen of the common cat's tail grass (*Phleum pratense*) weigh just one gram (0.035 ounces)! The whole process is very wasteful, since only a tiny proportion of the pollen produced will reach a receptive stigma, though it is clearly effective and not quite as chancy as it might seem. The flowers of wind-pollinated plants are usually readily recognisable by their reduced petals (or even the complete lack of petals), their long stamens and large stigmas, and their dull colour. The two main groups of wind-pollinated plants are trees (all common British trees except the limes are wind-pollinated) and the grasses, though plantains, meadow-rue, and various other related species are also wind-pollinated. Trees assist the process in various ways: they all flower early (except the limes) before the leaves open, so that pollen movement is not inhibited; the male and female flowers are often separated to ensure that there is no physical barrier to cross-pollination; the flowers are borne high up so that the pollen has a good start in its airborne life; and trees live in groups (woods). The pollen of trees is extremely widely dispersed, and pollen counts in mid-Atlantic have revealed the presence of significant quantities of pollen of birches, beech, oak, alder, pines, plantains and several grasses! It is hardly surprising that most female flowers are pollinated, though of course most of the pollen is wasted.

The other great group of wind-pollinated plants, the grasses, generally produce their flowers in high summer. Masses of pollen is produced from long dangling stamens, and it is this that is the primary cause of summer hay

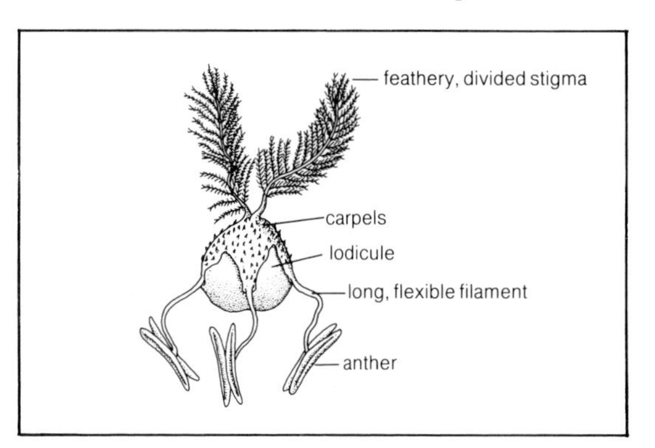

A typical grass flower.

fever. The female parts of the flower are usually surmounted by divided feathery stigmas, presenting the maximum surface area to catch drifting pollen. As with the trees, the method is clearly successful, and the grasses are amongst the most dominant plants of the flower world.

WATER POLLINATION

A small group of plants living on, in or close to water use water as a mechanism for transporting their pollen and achieving cross-pollination, though this is not common. A few flowering plants, such as the eelgrasses (*Zostera* spp.), live most of their lives submerged by the sea and achieve pollination completely under water.

Some aquatics, like *Gallisneria*, release tiny male flowers, from beneath the water, which float to the surface and release their pollen onto the female flowers. Others simply release their pollen onto the water surface and it is gradually and haphazardly transferred to any female flower of the same species at about water level. A few species produce their pollen in strands so that it has more chance of catching onto a female stigma as it moves along in the current.

SELF-POLLINATION

Some plants are self-pollinated, either habitually or, as we have seen, as a fail-safe mechanism when attempts at cross-pollination have failed. Some plants, like violets, produce 'normal' flowers in spring for cross-pollination, though subject to the vagaries of spring weather, but they also produce separate flowers later designed solely for self-pollination. These flowers would not be recognised as violet flowers, since they are green and never actually open, but they contain all the necessary parts (male and female) and simply pollinate themselves internally. Such flowers that never open are

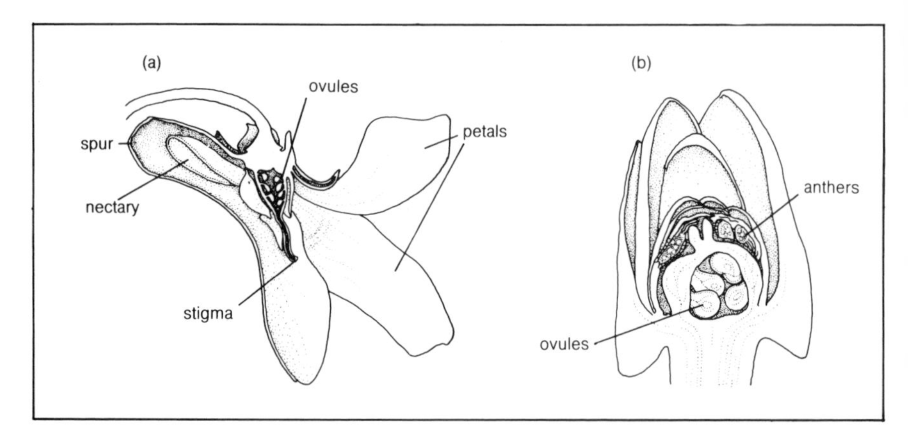

(a) Normal flower of violet, in section. It is pollinated by bees and other insects. (b) Cleistogamous flower of violet (in section), which produces seeds by self-pollination without ever opening.

known as cleistogamous flowers, and are produced by a wide range of plants either after the normal flowers, at the same time as the normal flowers, as in the henbit dead-nettle (*Lamium amplexicaule*), or even instead of the normal flowers, as in some helleborine orchids (*Epipactis* spp.).

The effect of self-pollination lies between cross-pollination and vegetative reproduction in terms of the variation it produces in the offspring; the genes of the plant can recombine in different ways, but no fresh genetic material is introduced.

Apomixis

A very few groups of plants, notably the dandelions (*Taraxacum* spp.) and hawkweeds (*Hieracium* spp.), produce flowers but frequently dispense with pollination altogether. The ovules in the carpels can simply develop into seeds without any fertilisation at all. This is, in effect, almost a vegetative form of reproduction, since no sexual fusion is involved and the offspring are all very similar to the parent. Because, in most groups, sexual reproduction takes place occasionally, the net result is that 'tribes' of similar plants differing only slightly from each other occur. This produces enormous problems for the plant taxonomist: sometimes the tribes are different enough to be called separate species, hence the enormous number of dandelion and hawkweed species recognised, and sometimes they are not quite different enough. Because all the plants in one tribe are so similar, it is tempting to classify them in some way, and they are often called microspecies; there are, for example, over 400 species of hawkweed (*Hieracium*) recognised in Britain alone! This process is known as apomixis.

Fertilisation

Pollination is simply the process of transferring the pollen to the stigma by whatever method is adopted. Once the pollen has reached the stigma it still has to germinate, grow down through the style to the ovary, and join with an ovule (egg) to form the first cell of a new seed and plant. The process of union of the male pollen cell and the female ovule is known as fertilisation.

When the pollen grains reach the stigma they are encouraged to grow, unless they are self-sterile, by a secretion containing sugars and hormones. The pollen grain pushes out a tube through a weak point in its outer covering, and this grows its way into the stigma and down into the style. The rate of growth of this pollen tube varies enormously: some reach their destination (the ovule) within a day of germinating, but in extreme cases they

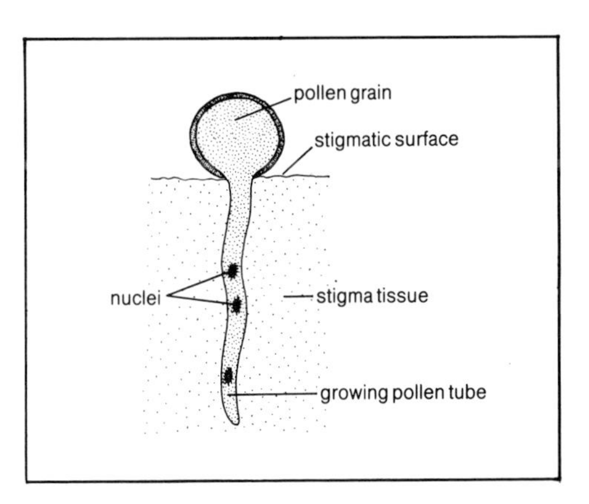

Pollen grain germinating on stigma.

may take a year. At the same time the nucleus of the pollen grain divides into two, though it still remains as one cell, and both nuclei pass into the growing tube. The leading nucleus divides again into two, and one of these fuses with the ovule to form the first cell of the new seed, while the other fuses with the nucleus of the embryo sac to produce eventually a food store for the germinating seed. It is interesting that pollen of the 'wrong' species will very rarely germinate on a stigma, and once the ovules have been fertilised even the 'right' pollen grains are no longer stimulated to germinate.

Division of the chromosomes

The cell for the seed which will develop into a new plant arises, as we have seen, from the union of the male nucleus from the pollen and the female

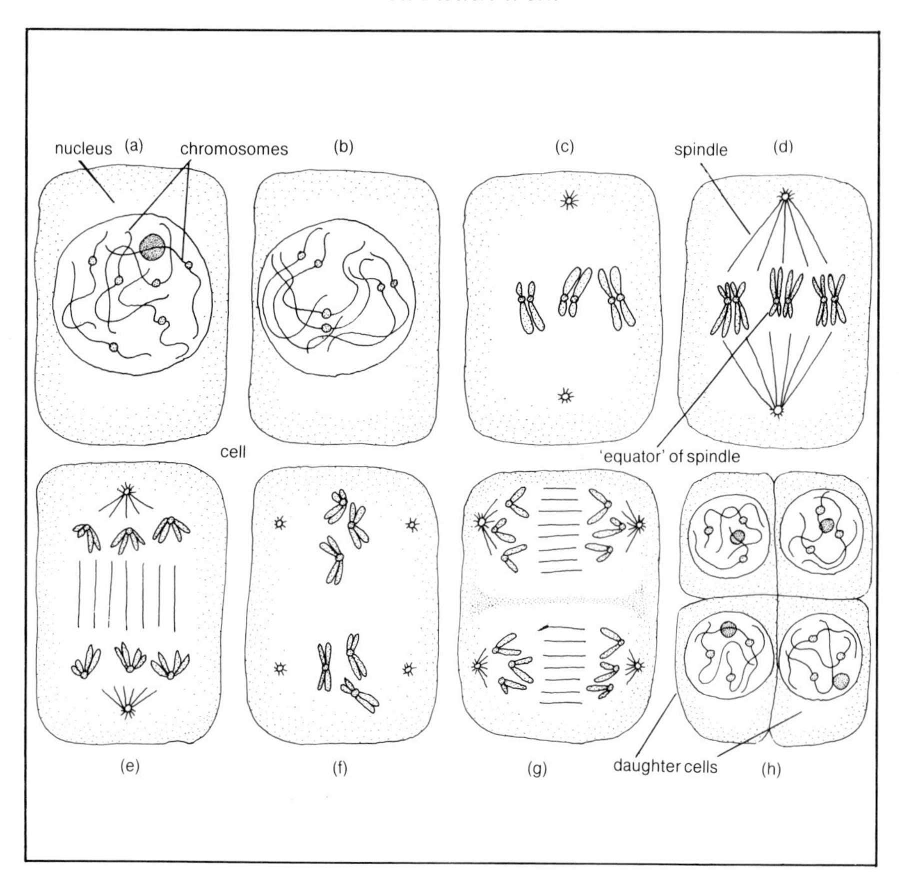

The stages of reduction division (meiosis), during which the number of chromosomes is halved. (a) Chromosomes become visible. (b) They shorten and begin to pair up. (c) They are short and clearly paired. (d) The pairs move towards the equator and the spindle forms. (e) The paired chromosomes split and move towards the poles. (f–h) Four new cells are formed, each with half the original number of chromosomes.

nucleus of the ovule to form a joint fertilised cell. So why does it not have twice as many chromosomes as either of its parents if each cell has brought a full complement of chromosomes with it? The answer lies in the method of formation of the reproductive cells by a particular process known as reduction division. It may be recalled that in normal cell division, such as takes place in growing points (*see* Chapter 3), the chromosomes replicate themselves and the halves split, one going into each new cell. In reduction division preceding sexual reproduction, each chromosome joins with a similar one and becomes linked to it. Then a process known as 'crossing over' occurs in which genetic material is exchanged between the two. When the central spindle forms, as in normal cell division, the two chromosomes are separated again, but they now consist of mixed chromosomes with genes from each of the two recombined. The separated chromosomes then move to opposite poles of the spindle and two new cells are formed, but each now has *half* the original number of chromosomes compared to the parent cell (and all the other cells in the plant) but these contain parts of all the chromosomes that were present in the original cell. These two cells divide again so that four haploid cells (i.e. with half the original number of chromosomes) are formed from the original 'mother' cell. In the case of pollen grains, all these four cells survive, but in the development of an ovule, three cells degenerate leaving the strongest as the haploid ovule.

When the male and female cells join at fertilisation, a nucleus containing the full complement of chromosomes is formed, but it has genes, and therefore characteristics, from each of its two parents. There are thus two ways in which variation is introduced: each pollen grain and each ovule has a slightly different chromosomal make-up from its fellows in the same plant because of the way in which the 'crossing over' of the chromosomes takes place; and then each of these pollen grains, if successful, recombines with an ovule of a different make-up. The possibilities are endless, and no two sexually-produced offspring are likely to be quite the same, providing enormous quantities of material for natural selection to work on. It can also be seen how the progeny of a self-pollinated plant will vary: the pollen grains will differ slightly from each other and their parents according to exactly how the chromosomes 'crossed over' and exchanged material, and so will the ovules. Nevertheless, they will, of course, only vary within narrow limits as no new genetic material can be introduced into the offspring, only variations of the old. Vegetative reproduction (described below), in which the new plants have exactly the same genetic constitution as their parents, provides no new material for natural selection to work on at all.

The fertilised ovule

Whatever method of pollination is adopted—insects, wind, or self-pollination—the net result is that each of the ovules becomes fertilised by union with the male nucleus from the pollen. The resulting combined cell is then the first cell of a seed which will ultimately develop into a new plant. The development of the seed from this single cell, and its subsequent dispersal and germination, are considered in Chapter 6.

VEGETATIVE REPRODUCTION

So far, we have only considered the ways in which flowering plants produce flowers and, as a result, produce seeds. This is the most widespread way in which flowering plants reproduce themselves, but many species have developed alternative non-sexual ways of reproducing themselves which may be used in addition to, or even instead of, the seed-producing sexual process. The essence of vegetative reproduction is that parts of the parent plant (propagules) are detached and give rise to new plants, though the variety of ways of achieving this are enormous. The advantages are that the risky processes of pollination, fertilisation and seed-production are bypassed, and the high-mortality stage of seedling germination may be completely avoided if the new plant starts life while still attached to its parent. This may have particular advantages under some conditions, though, as we have seen, it has evolutionary consequences by reducing the amount of natural variation within the species. Perhaps the best-known examples involve modifications of the stem, such as tubers, runners, stolons, rhizomes and so on. Potatoes are stem tubers formed underground by a swelling of the stem, utilising the existing tissues and incorporating a number of buds (the 'eyes'). As every gardener knows, a single potato plant can produce many tubers, and each of

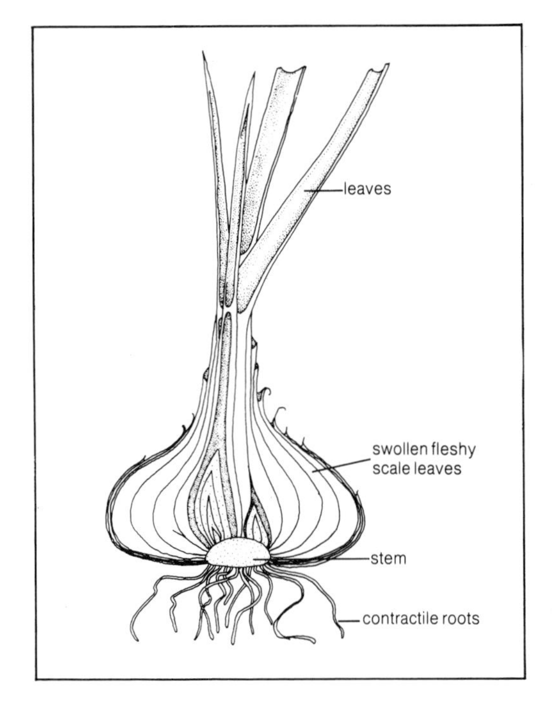

Section of developing bulb.

*Some plants reproduce vegetatively by producing short 'offsets', each of which can grow into a new plant. The photograph shows the original plant of a house leek (*Sempervivum tectorum*) surrounded by a ring of offsets.*

these can give rise to a new potato plant identical to its parent.

Bulbs and corms are also familiar examples, because of the ease with which they can be lifted and transplanted in their dormant stage. In natural conditions they act essentially as a storage organ, but vegetative reproduction can take place through a new bulb or corm reproducing itself by growing additional new ones. In fact, the two are of rather different origin: bulbs consist of a base plate, a modified part of the stem, on top of which develop a large number of sheathing non-photosynthesising leaves, surrounded by a papery bract. The leaves store food materials, and these provide for the growing shoot which emerges from the top when conditions are suitable. Onions are good examples of bulbs, and shallots are good examples of bulbs dividing to produce new ones. Corms, in contrast, consist solely of the swollen base of a stem covered in a membranous protective sheath. Each year, as the previous year's reserves have been exhausted, the

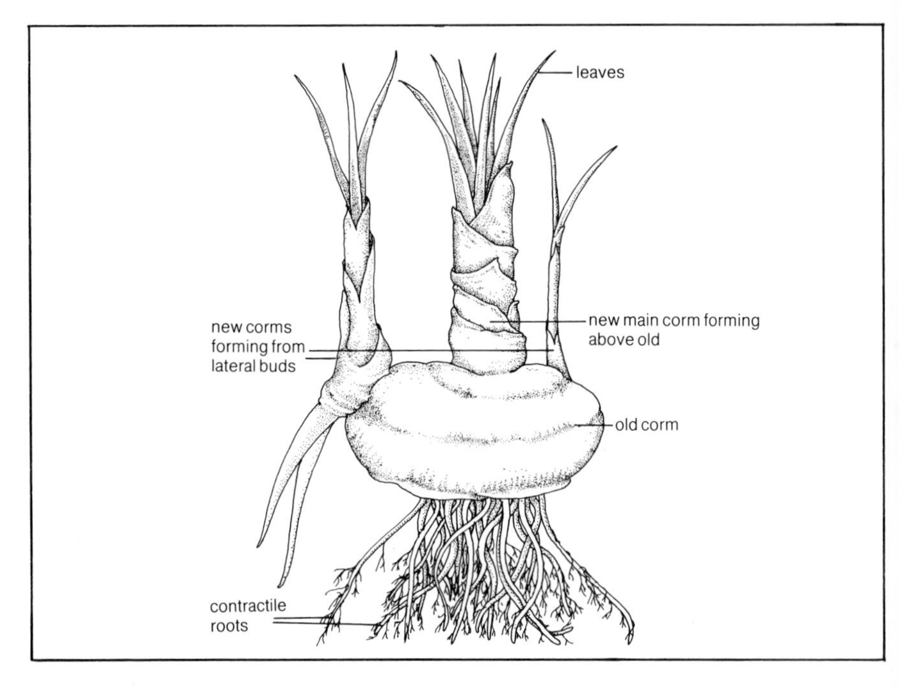

Reproducing corm of crocus.

next year's reserves are laid down as a new corm above the old one. The stem may branch and produce new corms which can then separate and produce new plants. At times, it is difficult to say whether a plant is simply becoming larger or actually reproducing: reproduction can only really be said to have taken place when the offspring leads an independent life.

Many plants produce horizontal stems which grow along the ground or below it and can give rise to new plants at intervals along their length. When they are produced above ground, they are known as runners, and familiar examples include strawberries and the creeping buttercup (*Ranunculus repens*). The plants thus produced can quickly become independent once rooted. Underground rhizomes are produced by such well-known and persistent weeds as couch grass (*Agropyron repens*) and bindweed (*Convolvulus arvensis* and *Calystegia sepium*), and this is the secret of their success. Extensive systems of underground rhizomes occur, and these can give rise to new plants at frequent intervals; even detached portions of rhizome, provided they contain a bud, can produce new plants and this renders them exceptionally difficult to eradicate, especially by rotary-plough cultivation which simply breaks up the rhizomes into numerous new potential plants. A

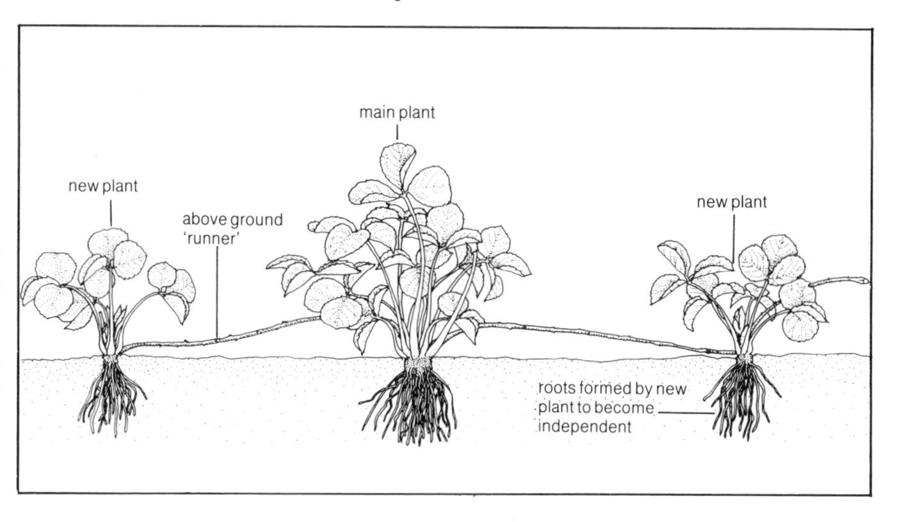

Diagrammatic representation of vegetative reproduction by strawberry plant.

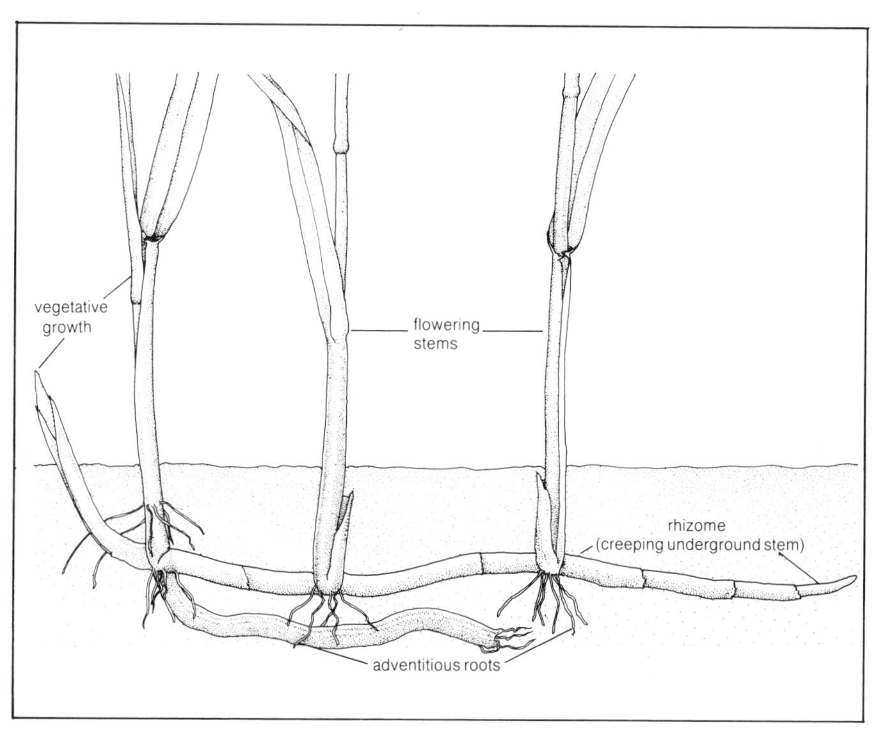

*Vegetative parts of couch grass (*Agropyron *sp.). This plant owes much of its success as a persistent weed to its creeping rhizomes.*

few plants, notably the house-leek (*Sempervivum tectorum*), produce very short stems with new plants, known as offsets, on the ends. The plants may root *in situ* or break off and establish somewhere else, wherever they end up.

Plants such as the garlic plants (*Allium* spp.) may produce vegetative propagules known as bulbils in place of flowers in the flowering head. These develop directly from the parent tissue, with no reduction division in their formation, but they can act in effect as large well-nourished seeds and be dispersed from the parent plant to grow into new plants in new locations.

5 · FRUITS, SEEDS AND SEEDLINGS

The joining of the male and female cells, after pollination, to form the fertilised egg is, in one sense, the beginning of a new generation, as the seed will develop and be dispersed finally to germinate and grow into a new plant. Before looking at this remarkable process and the vast variety of ways that plants have evolved to disperse themselves, it is worth clarifying a few definitions since the general everyday use of the terms 'fruit' and 'seed' often differs from their botanical meaning. The everyday use of 'fruit' is as a juicy, edible and usually sweet plant structure such as an orange, apple, or grape; at the same time true fruits like tomatoes, cucumbers, beans, marrows and many others are not thought of as fruits at all, either because they are not sweet or because they are eaten as vegetables.

Botanically, a fruit is the mature female part or parts of the flower (the carpels) which may or may not include other parts of the flower developed for specific purposes. The seed is the ripened ovule contained within the fruit. For example, the grape fruit is the whole structure including the fleshy edible section, while the 'pips' are the seeds, and the same applies to oranges, apples and many other edible fruits. In cherries or plums the whole structure is the fruit and the kernel inside the 'stone' is the single seed contained within it. In examples like this the difference is clear enough since the fleshy parts of the fruit are so distinct from the seeds; in other cases the distinction is less clear especially where only one seed lies inside a barely-modified carpel. For example, the fruits of the buttercup are the individual 'knobbles' on the green spiky cluster which develops after flowering; each of these contains one seed, taking up most of the space, and the two behave in effect as one and the same. Occasionally the origins of what we call the fruit are even less clear; for example, the vast bulk of the strawberry fruit is made

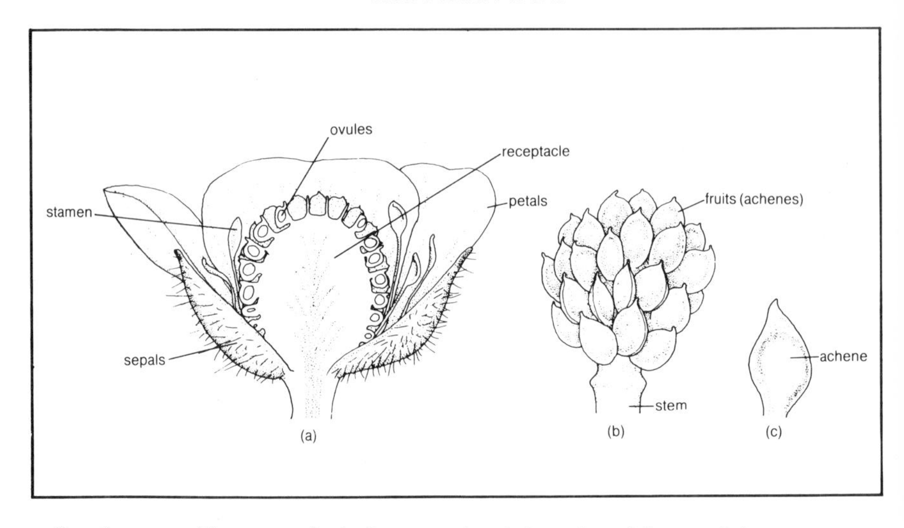

Development of buttercup fruits from ovaries: (a) section of flower; (b) developed fruits; (c) single fruit.

up of the swollen base of the flower or top of the stem, the receptacle, and the tiny seed-like structures lying on its surface are the true fruits, each containing one seed. The difference between the fruit and the seeds contained within it will become clearer as we look at the development of the fertilised ovule and the pattern which leads to the development of a mature fruit containing seeds.

DEVELOPMENT OF SEED AND FRUIT

At fertilisation the embryo is formed of one cell caused by the union of the pollen nucleus and the female ovule (*see* page 69). This cell divides first into two equal cells, then, from this point onwards, the developing embryo begins to differentiate into separate stem, root and leaf shoot components. One end of the embryo forms a very large cell which will ultimately produce the whole root system of the plant, while at the other end the structures known as the seed leaves (cotyledons) become apparent. Most plants (the dicotyledonous plants, *see* Chapter 1) have two such seed leaves, with the shoot which will produce the whole of the above-ground part of the plant contained between them. All other plants (the monocotyledonous plants) have only one seed leaf with the shoot's growing point held to one side of it. Normally these seed leaves emerge above ground after the seed germinates and provide the first source of manufactured food, through photosynthesis, to enable the plant to survive. The seed leaves are usually quite different in form from the leaves that develop later and which give each plant its characteristic appearance. In a few plants, such as the garden pea, the first true leaves are well developed within the seed, though in most cases the seed simply contains the seed leaves and the shoot.

The remainder of the developing ovule, still within the immature 'seed', is taken up by a mass of tissue which contains stored foods translocated into it by the adult plant. This is known as the endosperm, and it acts as a food source for the developing embryo. The provisioning of each developing ovule with food can pose a considerable strain on the plant, and this is one reason why fewer, smaller seeds are produced in poor conditions. At the same time, it is partly this provisioning of the endosperm with food that makes some seeds such an attractive food source for humans (and many other animals); such seeds include the staple foods like wheat, barley, and rice as well as larger seeds like castor-oil beans and coconuts. All are examples of seeds with well-provisioned endosperms.

Other seeds, instead of having a well-provisioned endosperm around the embryo plant, store most of the food for the germinating plant in the embryo itself within the seed leaves. If you look at a bean or pea seed, you can clearly see that the two swollen seed leaves take up most of the space in the seed, and it is these that form the bulk of the food since they contain all the stored carbohydrates and proteins. These are then gradually used up as the seed germinates, until the leaves are above ground and able to start producing food for themselves.

A third category of plants produces seeds that have virtually no food reserves at all, and only contain a tiny embryo. Here, the strategy is quite different and most of the plant's energies go into producing large numbers of tiny seeds rather than a few large ones. Naturally, the success rate is much lower, but the seeds will be very widely dispersed and will reach many potential new colonisation sites. Often such a strategy is associated with a parasitic way of life (e.g. the broomrapes (*Orobanche* spp.) which live on the roots of other plants), such plants needing to disperse their seeds widely to have a chance of finding a new 'host', or the seeds are heavily dependent on making an association with a soil-living fungus (called a mycorrhizal association) before they can successfully germinate. This reduces their need for an internal food source, but also adds to their requirement for wide distribution in the hope of encountering a fungus. The numbers of such seeds may be enormous—over a million seeds can be produced by one plant! However it stores its food, each seed has around it a tough coat to prevent damage and excessive absorption of water, and is known as the testa;

The strawberry is not really a fruit but a vastly swollen stem in which numerous tiny single-seeded fruits are embedded.

examples are the 'skin' of a single broad bean and the dry brown papery covering around a single peanut or hazelnut.

At the same time as the ovule is developing inside the carpel into a seed, so the remainder of the carpel and the rest of the flower are undergoing changes. Normally, after adequate fertilisation of the ovules has taken place, the sepals (calyx), petals and male parts serve no further purpose, and they wither away and die. In other cases, as we shall see, they develop or persist and play some part in the dispersal of the fruit. In a few cases they persist but play no obvious part, such as the green ring of sepals that remains attached to tomato or strawberry fruits.

The most important development, which takes place in virtually all plants, is the development of the carpel (or ovary) surrounding the ovule to form the fruit. As this develops it becomes known as the pericarp of the fruit, and in most cases it forms the remainder of the fruit other than the seeds. For example, a pea pod is a fruit, and it consists of the 'pod' itself which is the pericarp developed from the original ovary, and the 'peas', each of which are seeds within their own protective testa (the skin) developed from the original ovules. In a plum fruit the ovary has developed into a pericarp with three distinct layers: the outer skin of the fruit (not to be confused with the testa of the seed), a fleshy layer below this (the flesh of the plum), and a hard horny layer inside this which forms the outer layer of the 'stone'; these three layers are developed from the ovary or carpel, and are therefore not part of the seed. Inside the stone lies the 'kernel' within its own thin protective testa, and it is this which is the true seed.

The number of seeds within each fruit varies enormously, depending on the number of ovules produced in the ovary, the number that become fertilised (this is very rarely all of them), and whether several carpels are joined together to form one fruit, e.g. in the capsule of a lily (*Lilium*) which consists of three united carpels. The number may vary from one, as in the example above (the plum) and in the cherry, almond and other similar fruits, through several in pea or bean pods, to enormous numbers; for example, over 40,000 seeds have been counted in one capsule of the common corn poppy (*Papaver rhoeas*) and some tropical orchids regularly contain a quarter of a million seeds in a capsule!

In a number of plants it is not only the developed ovary that makes up the fruit. We have already considered the example of the strawberry where the true fruits are the tiny 'pips' lying on the surface of the massively swollen flower base. In the apple fruit, and the broadly similar rosehip, the ovary only makes up the central portion of the fruit (more or less corresponding to

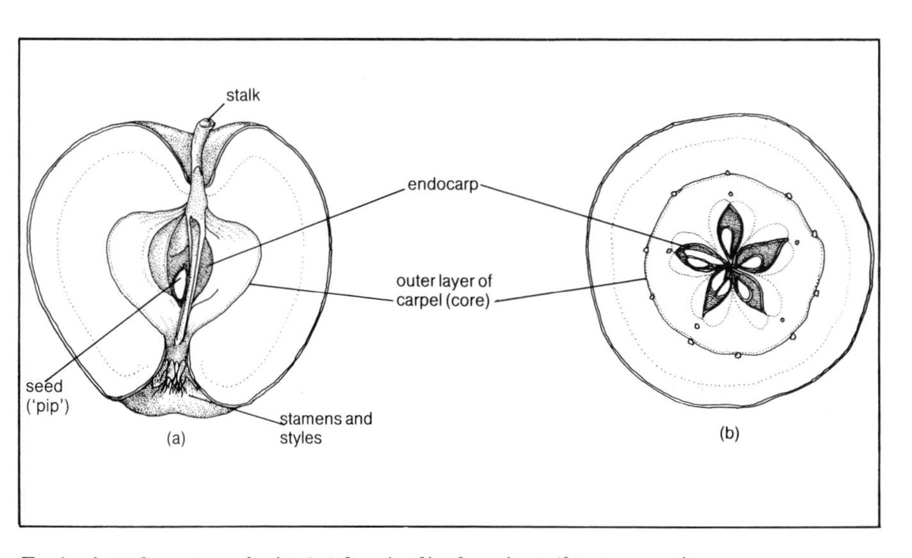

Fruit of apple, a pome fruit: (a) longitudinal section; (b) cross-section.

the apple core that we leave), and the rest is made up of the receptacle (the base of the flower) which grows up around the developing ovary. The boundary between the swollen receptacle and the true ovary lies in the harder pithy area around the seeds which is slightly less palatable. In this case the swollen receptacle contributes greatly to the attractiveness of the fruit and renders the seeds more likely to be dispersed, though of course the size of modern apples is an artefact of plant breeding by humans (*see* Chapter 8).

In other cases, a different part of the flower plays a significant part in the developed fruit: for example the fruits of dandelion (*Taraxacum* spp.) and many of its relatives in the daisy family (Compositae) have a miniature parachute to aid dispersal, and this is developed from the ring of sepals (calyx) in the original flower, whereas in the old man's beard (*Clematis vitalba*) or the pasque flower (*Pulsatilla vulgaris*), both in the buttercup family, it is the style that develops into the 'parachute' to aid dispersal.

Because fruits vary so widely in their form, origin, number of ovules and so on, botanists have found it convenient to classify them into the following types: (a) simple fruits, derived from a single ovary; (b) aggregate fruits, derived from a number of ovaries belonging to a single flower and sitting on the same receptacle (good examples are the blackberry and the raspberry which consist of a conglomeration of 10 to 20 separate 'fruitlets'); and (c) multiple fruits, derived from a number of ovaries from several flowers which have more or less grown together into a single unit, e.g. the pineapple, mulberry or fig, all of which consist of the ovaries of numerous flowers joined together in different ways.

Each of these three main types has been subdivided further, especially the simple fruits of which there are numerous types. There is little need to look at all the different sorts, though a few well-known examples may be of interest.

Examples of simple fruits include capsules, which are dry splitting fruit derived from several fused carpels; they split (dehisce) in several ways, including lengthwise, as in *Iris* capsules, by pores at the top, as in poppies (*Papaver* spp.), or by a transverse lid, e.g. plantains (*Plantago* spp.). Achenes are single-seeded non-fleshy fruits in which the seed forms most of the bulk, and indeed the whole may often be called a seed, e.g. sunflower 'seeds', buckwheat, buttercup fruits (*Ranunculus* spp.) and the single-seeded fruits on the surface of the strawberry. A samara is a dry non-splitting fruit with one or two seeds characterised by an outgrowth of the ovary wall forming a 'wing', as in sycamores and maples (*Acer* spp.) or ashes (*Fraxinus excelsior*). A berry is a fleshy, non-splitting fruit derived from a compound ovary: seeds from several fused ovaries are embedded together

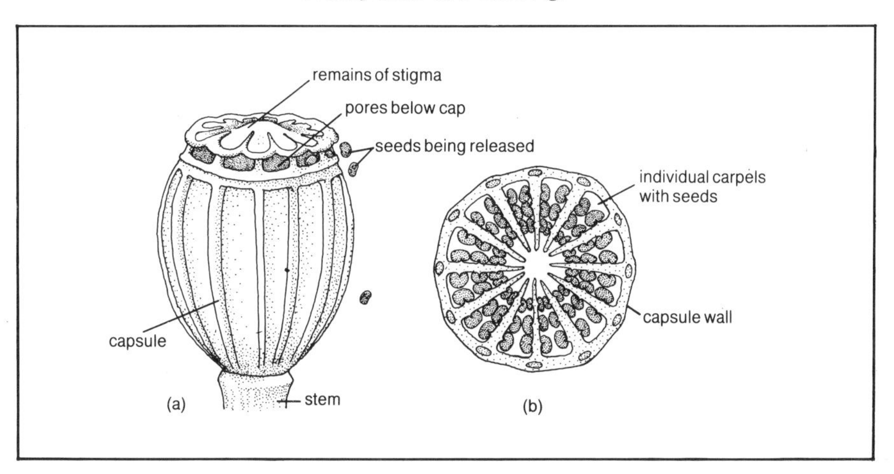

Fruit (capsule) of poppy: (a) capsule shedding seeds; (b) cross-section.

in the flesh, as in the tomato or blackcurrant. Most fruits with 'berry' in their name—blackberry, strawberry etc—are quite different forms of fruit botanically.

Aggregate fruits derived from a number of separate ovaries in one flower can usually be classified according to the type of simple fruit that makes them up: for example, the strawberry is a swollen receptacle covered with achenes; a blackberry or dewberry (*Rubus* spp.) is a collection of small drupes (single-seeded fleshy fruits like plums or cherries) and so on.

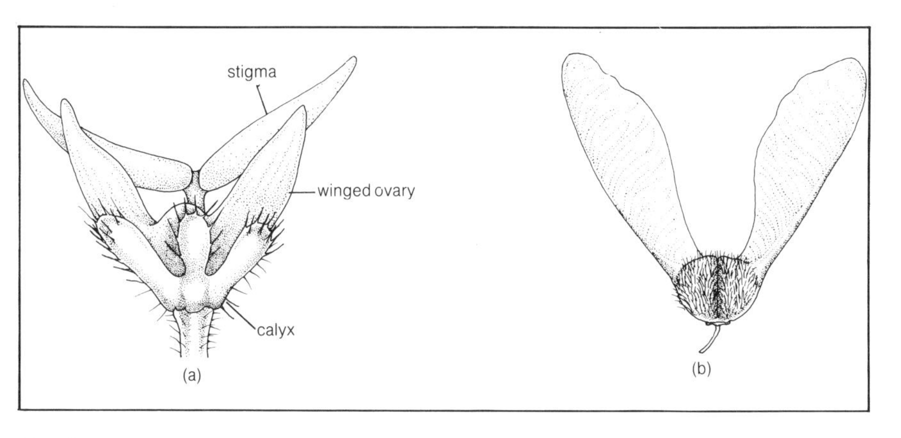

Development of samara (winged key fruit) of maple: (a) flower; (b) mature fruit.

SEED DISPERSAL

Each fruit type, whatever name is given to it, reflects a different strategy for dispersing the seeds and giving them a good start in life. It is very important for plants to be able to move around as conditions become unsuitable, as new opportunities arise, and to prevent overcrowding. A great variety of mechanisms for dispersing seeds has evolved, using wind, water, mammals, birds, insects or other means reflecting whatever conditions the plant has evolved in, the means available to it, and the requirements of its life cycle. For example, 'opportunist' plants that colonise newly-bared ground, such as the fireweed or rosebay willowherb (*Chamaenerion angustifolium*) and groundsel (*Senecio vulgaris*), require seeds that are highly mobile to find the new habitats before other species colonise them and make them unsuitable.

Wind dispersal

Many seeds are dispersed by the wind. At its simplest, this involves exceedingly light seeds which are so tiny that they behave like dust and travel long distances before settling back to earth. There are many known instances of orchids, for example, colonising new sites 50 or 100 miles (80 or 160 kilometres) from their nearest locality, and it is their incredibly light seeds that allow them to do this, though of course most of the production is completely wasted, as it never lands anywhere suitable.

Other plants produce seeds or fruits which are much heavier, and contain substantial food reserves, yet are made sufficiently buoyant by the addition of wings, parachutes or other buoyancy mechanisms. A well-known example is the double samaras of the sycamore (*Acer pseudoplatanus*) or the field maple (*Acer campestre*) where the fruit is elongated to one side into a long 'propellor' blade. This off-centre centre of gravity causes the fruit to spin and drift, and the final distance of travel is quite long because of the high starting point. Other trees, such as elm or ash, employ similar devices, whereas the fruits of the lime (*Tilia* spp.), though unwinged themselves, are

---

Above: The fruits or 'keys' of the wych elm (Ulmus glabra *) are an example of a winged fruit (samara) that is dispersed by wind currents.*

Below: The seeds of the corn sow-thistle (Sonchus arvensis *) have a feathery parachute structure attached to them to aid dispersal by the wind. The seeds can be seen below the parachutes.*

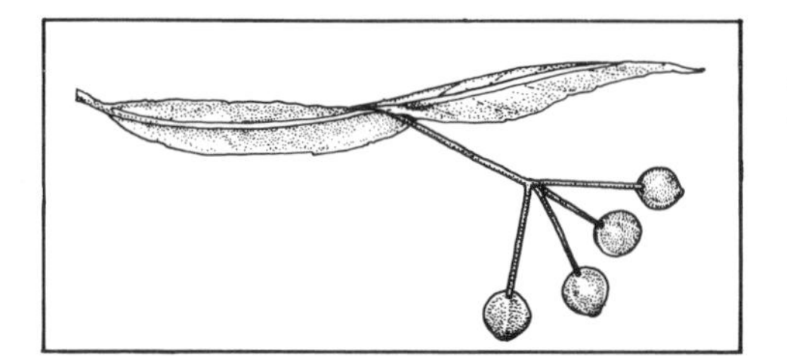

*The leaf bract of lime (*Tilia *spp.) detaches with the fruits and aids wind dispersal.*

attached in groups to a wing-like bract which aids their dispersal. These mechanisms do not serve to carry the seeds for very long distances, but at least they carry a substantial proportion out from under the canopy of the parent.

Many plants from a wide range of families employ some form of feathery device to catch the wind and effectively lighten the seed—dandelions, thistles, ragworts, rosebay willowherb (*Chamaenerion angustifolium*) and old man's beard (*Clematis vitalba*) are all familiar examples. Many fruits of the daisy family, the Compositae, have beautifully balanced 'parachutes' with the fruit hanging directly below them, and are able to travel long distances despite the relatively heavy seeds. One of the best examples with large seeds and a 'parachute' over half an inch (13 millimetres) across is the goat's beard (*Tragopogon pratense*). In other cases, such as thistles or rosebay willowherb, the mechanism is less finely symmetrical yet equally effective. Both old man's beard and the beautiful pasque flower (*Pulsatilla vulgaris*) have elongated styles (the stalk that bears the stigma) which become covered in downy hairs and act as additional wings to carry the seed aloft.

Animal dispersal

Other plants make use of animals to disperse their seeds, and many of these, such as 'burs' or goosegrass, are well-known to children for their ability to stick to animals, particularly humans! The whole goosegrass or cleavers plant (*Galium aparine*) is covered with sticky hooks, and as the plant ages it may come apart so that whole sections, with their fruits, may be dispersed on passing animals. Burdocks (*Arctium* spp.) produce burs, which are sticky balls with long hooked spines which attach strongly to any fur or material; eventually the hooks break and the seeds fall to the ground in a new place. Similar strategies are adopted by the water avens (*Geum rivale*) and the sharp barbs of Spanish needle (*Bidens bipinnatus*). All such plants depend on the regular presence of mammals at the right time of year, and their distribution naturally tends to follow the pattern of movement of the animals; hence they are normally plants of path edges and clearings.

The 'burs' of burdock are covered with numerous hooked spines which attach to passing animals, aiding dispersal of the fruits.

Other species make use of animals, and especially birds, by producing highly attractive palatable fruits with very resistant seeds. Hawthorns (*Crataegus* spp.), rowans (*Sorbus aucuparia*) and blackberries (*Rubus* spp.) all use this mechanism, in which the fruits are eaten by birds and the seeds are later excreted at whatever point the bird has reached at that time. This mechanism also provides a small amount of 'fertiliser' which may help the seedling on its way. A similar, but rather more specialised, adaptation is the well-known dispersal of mistletoe (*Viscum album*) by thrushes and black-birds which eat the fleshy berries and then wipe the sticky seeds from their beaks on a branch, often slightly wounding the tree and providing a perfect germination site for this parasitic plant.

Birds are probably made use of in a different way by plants which produce seeds that become sticky when wet. Such seeds are picked up on the feet of birds and then may be transported for considerable distances, especially by migrating birds, and deposited. There is little information on this, and it is clearly a chancy business, yet many examples of long-range dispersal can only be satisfactorily explained in this way. Mammals may be involved too, not to mention walkers' boots, car tyres and even botanists' polythene bags.

A few groups of plants provide special attractions for insects, especially ants which are attracted to oily and sugary substances and tend to forage widely. The seeds of violets (*Viola* spp.) and milk-worts (*Polygala* spp.) have special oily bodies known as elaiosomes whose sole function seems to be to attract ants to disperse the seeds. Not surprisingly, the most likely result of this is that the seeds end up in ants' nests, so they must be able to germinate under such conditions.

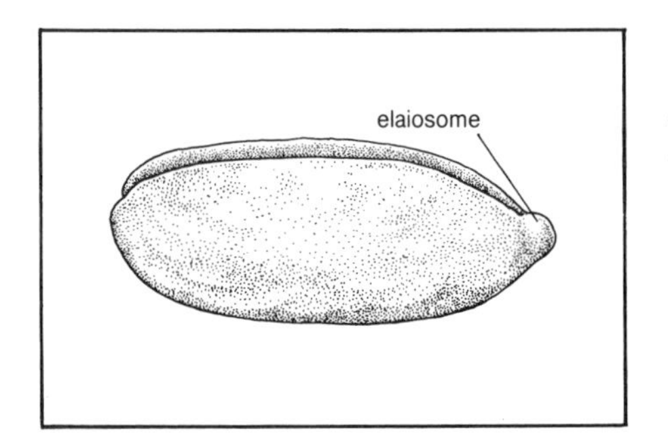

Seed with oil-producing body (elaiosome), which attracts ants and aids the dispersal of the seed by the ants.

Water dispersal

Many plants that grow in or near water make use of water for dispersing their seeds. The structures are not usually elaborate; all that is required is some buoyant tissue, usually provided by air spaces. This method is used by such widely-differing plants as alder trees (*Alnus* spp.), water lilies (*Nymphaea* spp.), many waterside sedges (*Carex* spp.) and gipsywort (*Lycopus europaeus*). At some stage the seeds come to rest on the bank or the bottom and will germinate there if conditions are right, as they often are.

Self-dispersal

A few plants have developed specialised methods of dispersing their seeds without the aid of any outside agency, mainly by some form of explosive scattering mechanism. Most, though not all, of such methods rely on the forces set up as differently-aligned tissues dry out and tend to contract in different directions. This tends to twist the pod or other fruit, and the tension is released when the tissue holding the seams finally gives way, releasing the contents at high speed. It is a common experience to walk through a gorse (*Ulex europaeus*) thicket on a hot day and hear the seed pods 'popping' with a crack as they release their seeds and scatter them in all directions.

Some sorrels (*Oxalis* spp.), especially *Oxalis corniculata* the garden weed, employ a slightly different method: the seeds are shot out of a slippery contractile coating, propelling them several feet, rather like cherry stones being squeezed between the fingers of small children!

The Himalayan plant policeman's helmet or Himalayan balsam (*Impatiens glandulifera*) is an attractive large pink-flowered annual that has invaded riversides and damp areas in many parts of the world. One of the secrets of its success is its seed dispersal mechanism; when ripe, the fleshy green capsules become swollen by water pressure from within, and any very slight touch or squeeze will cause them to 'explode', scattering the seeds in all directions in a most impressive way.

DORMANCY

Most seeds, however they are dispersed, eventually come to rest in a place that may be suitable for germination. Yet the seeds rarely germinate immediately, but undergo a period of inactivity or dormancy. This phenomenon has very significant ecological advantages, especially in the temperate regions of the world, and most plants use it to some extent.

The majority of plants produce their seeds at the end of the favourable growing season, making use of as long a growing period as possible to reproduce and ensure that all seeds are well-provisioned. This means that the release of the seeds coincides with the onset of an unfavourable period, i.e. winter, and it would not be advantageous to the plants to germinate then (a few plants do, though these are either rare, i.e. not successful, or have some special mechanism for overcoming the problem). Thus, most plants have mechanisms of some sort or other which prevent the seed from germinating immediately, and which allow it to do so either after a certain period of time has elapsed or when certain conditions or combinations of conditions prevail. At the same time, many plants employ a form of 'staggered' dormancy such that the seeds respond slightly differently and germinate at different times; a proportion may not germinate at all in the first season after release, but may remain dormant until the second or even third season. This helps plants to overcome any difficulties brought about by a very poor growing season or natural disaster, and allows a proportion of the population to continue to the next year. Another type of dormancy ensures that seeds only germinate when a certain periodic event occurs, such as desert plants germinating after storms, or plants that only germinate after a forest fire (giving them ideal growth conditions).

Mechanisms of dormancy

Plants achieve this ability in different ways. Many seeds or fruits contain a water-soluble germination inhibitor which prevents all germination until it has almost completely washed out or broken down. For example, tomatoes contain a highly effective germination inhibitor which can prevent germination of its own seeds, and those of other plants, even when diluted 25 times. If the seeds are separated and washed they will germinate rapidly, but normally the substance disappears over a period of months. The seeds of sugar beet, and many other plants, contain a similar water-soluble inhibitor. Other plants make use of volatile inhibitors which gradually evaporate, eventually releasing the seed from dormancy. Such mechanisms tend to prevent the seeds from germinating for a specific time, and they are therefore geared towards regular seasonal changes.

A similar effect is achieved by seeds that have a very tough impermeable outer coat. Such a coat prevents the entry of water, and sometimes atmospheric gases, so that no germination will take place. However, after a winter in the soil with frosts, abrasion by soil particles, alternate drying and wetting, and bacterial and fungal attack, the seed coat loses its impermeability and becomes able to imbibe water. A very fine balance has to be achieved, of course, to prevent rotting before conditions are right and yet ensure that the seeds are ready when conditions do become suitable, and only a proportion of seeds achieve this balance. The process can be artificially hastened by abrasion or puncturing of the outer skin, though conversely a seed may remain dormant for longer if stored in cool dry conditions where no attack on the seed coat is taking place.

Other seeds, obviously from plants in temperate regions, may require a period of cold before they will germinate. This may involve temperatures below freezing for some species, while for others a period of low temperatures above freezing may suffice; for example, in a test, seeds of the common penny-cress (*Thlaspi arvense*) did not germinate at all when kept at a constant high temperature, while 18 per cent germinated after two weeks at 3° Celsius and 100 per cent germinated after two months at 3° Celsius.

Such mechanisms prevent seeds from germinating in autumn when conditions are often quite similar to those in spring, but would result in failure if the seeds did germinate. It allows the seed to 'detect' when winter has passed, though the unpredictable winters of many West European countries can confuse the mechanism with mild spells in the middle of January. Some seeds require a combination of factors and may, for example, need light after a period of cold. Such requirements relate to the life cycle of the plant, and a light requirement, for example, is most often found in weeds of cultivated ground which would die if they germinated deep down in the soil, but can succeed when they are re-exposed by cultivation.

A few plants, notably Australian plants of the Protaeceae family, need a period of baking by fire to achieve germination. In some cases this simply forces the cones to open and release their seeds, while in others it actually breaks the seed's dormancy. Germination then coincides with a period after a forest fire when nutrient levels are high and competition levels low, giving ideal conditions for those that survived the fire.

Plants that grow in tropical conditions where there are few seasonal changes in temperature or rainfall tend not to exhibit any dormancy at all as there is no necessity for it. Such seeds are often very short-lived.

After the dormancy period, whatever its length, the seed must still be 'viable' in order to germinate.

SEED VIABILITY

Coupled with the ability to remain dormant is, of course, the ability of seeds to survive for long periods, and the length of time that a seed can remain alive and capable of germinating is known as its viability. Seed viability varies enormously from a few days to many years, though a great deal depends on the conditions that the seeds are in. In normal conditions in the wild, the seeds of willows (*Salix* spp.) and poplars (*Populus* spp.) only remain viable for a few days whereas the seeds of many weeds can remain viable, when buried in the soil, for at least 40 years. Seeds of the common arable weed charlock (*Sinapis arvensis*) have been found to be viable after at least 25 years, and seeds of the hedge mustard (*Sisymbrium officinale*) have remained viable for at least 40 years. The longest known periods of viability have been recorded in the seeds of the sacred lotus (*Nelumbo nucifera*) which are known to have survived and remained viable for about 120 years. The stories of seeds germinating after removal from Bronze Age archaeological sites or the tombs of the Pharaohs all seem to be without foundation.

It has been found, however, that most seeds can remain viable for a very long time indeed if their optimum storage conditions are found and then maintained. Most seeds survive best in cool, dry conditions, though there are exceptions, and other seeds, such as those of the semi-parasitic yellow rattles (*Rhinanthus* spp.) and louseworts (*Pedicularis* spp.) soon lose their viability if they dry out. There are now many 'seed banks' set up at scientific institutions throughout the world which attempt to store representative seed samples of economically or scientifically significant plants so that a source of these plants is always available.

GERMINATION

Of the many seeds produced by a plant, only a proportion will survive predation, rotting, falling in the wrong place or any of the many other natural hazards besetting a seed. Those that do survive will, sooner or later, germinate to begin to form a new plant. While the seed is dormant, all its processes are slowed right down so that few of the precious food resources are wasted though the embryo is still alive. When the dormancy is broken and conditions are right, the seed rapidly takes in water and the respiration rate rises back to normal as cells start to grow and divide. This means that the seed requires oxygen to fuel the respiration, and the temperature must be high enough to sustain growth.

As soon as the cells have all taken in water and become fully active again, the embryo starts to grow. The area of greatest growth at first is the root initials, and the young root soon pushes its way out through the seed coat. This anchors the developing seedling and provides a secure source of water and mineral salts. At this stage, the shoot (known as the plumule) begins to develop and emerges from the split seed. The part of the shoot above the seed leaves is going to grow into the whole of the rest of the plant, yet it has to push its way through several inches of soil which may be tightly packed, or stony, and it must not be badly damaged during this journey or the whole plant will be malformed. Most plants solve this problem by producing greater growth on one side of the stem than the other so that a 'loop' is formed, and the point of the loop pushes its way through the soil, pulling the all-important plumule behind it. Anyone who has watched seedlings emerge will have noticed the loops of stem that are often the first parts to appear above ground.

The seed leaves, or cotyledons, can play two different roles at this stage.

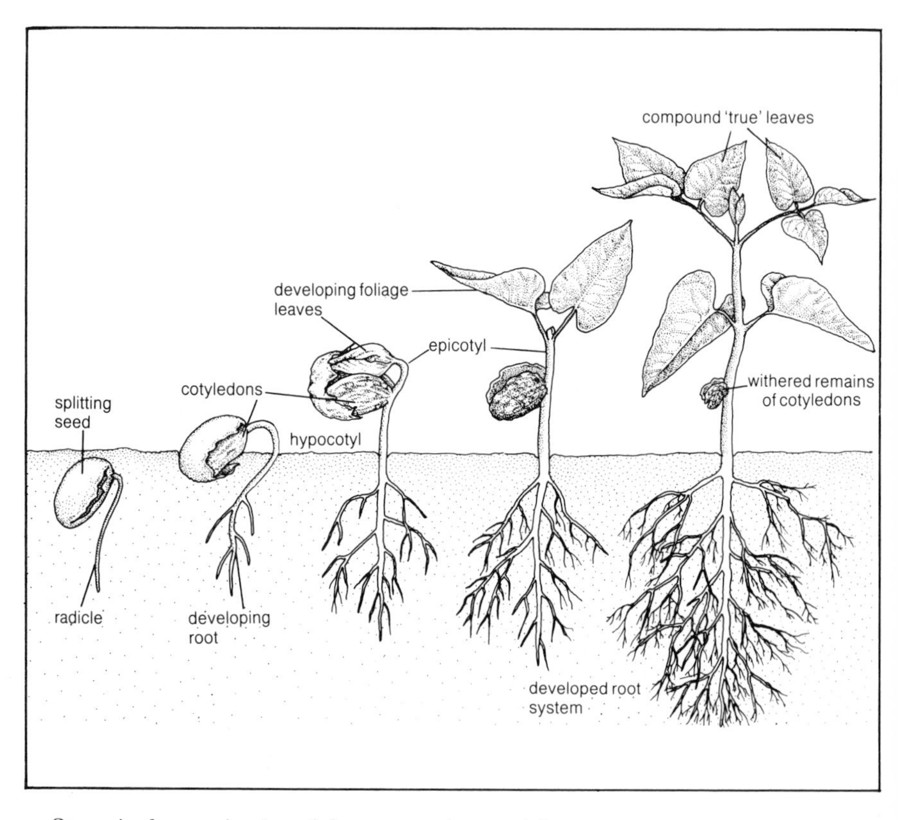

Stages in the germination of the common (runner) bean.

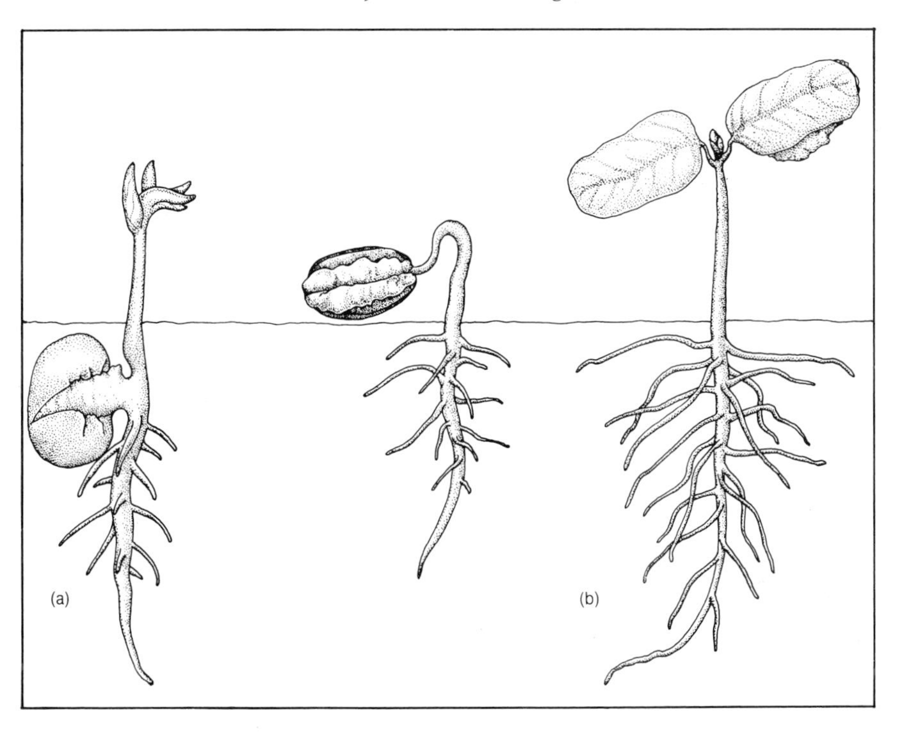

Different types of germination: (a) hypogeal germination, in which the cotyledons remain below ground; (b) epigeal germination, in which the cotyledons are raised above the ground.

As we have seen, they may or may not act as storage organs, and this is reflected in their subsequent behaviour. Generally, if they are the food storage organs, they remain below ground providing food for the developing plumule (and withering as they do so) so that the first leaves to appear above ground are true leaves, resembling the leaves of the adult plant. This is known as hypogeal germination. In many other plants the stem grows at a different point, and the seed leaves, together with the plumule and often the remains of the seed coat, are pushed above the ground so that the first leaves to appear are the seed leaves. This is known as epigeal germination. The seed leaves are generally recognisable since they occur as a pair (in dicotyledonous plants, at least) which are usually very regular and symmetrical and with few indentations. They frequently bear little resemblance to the leaves of the adult plant. Mustard and cress seedlings exhibit epigeal germination, and indeed most people know nothing of the adult plant as the seedlings are eaten before they reach the stage of developing true leaves.

Whichever form of leaf emerges first (seed leaves or true leaves) they are immediately able to start photosynthesising (*see* Chapter 3) and from then on the seedling is independent of any food reserves carried in the seed which by now will probably be exhausted.

It can be readily seen how important the food reserves in the seed are as they have to fuel all the root and shoot growth until the first green leaves are expanded above ground. Most seeds do not contain enough food to allow repairs to damaged tissues, and if something goes wrong the seedling will probably die. A few seeds with particularly large reserves, such as acorns, can provide for a new shoot if the first one fails, but most seeds cannot.

The stages from the swelling of the seed to the emergence of the first leaves are known as germination, and from then on the plant enters the establishment phase, as the growing plant consolidates its position. The speed at which this takes place depends largely on the strategy of the plant, affected by environmental factors, and some plants (usually 'weeds') will develop to the stage of flowering in a few weeks whereas others may take many years. These different strategies are examined in more detail in the next chapter.

6 · THE LIFE CYCLE

Throughout the year, plants undergo many changes: some lose their leaves but remain dormant above the ground; some disappear below ground; others die altogether. Some plants live for a few months, some for two years, and others for thousands of years. Some plants produce hundreds of thousands of seeds, others produce only a few, and some produce none at all. Yet, left to their own devices, very few plants become extinct, and all these different strategies are successful to a greater or lesser extent.

So, is there any pattern to these life cycles? How long do plants live, and do all individuals of the same species live for the same length of time? The grouping of plants into annuals, perennials and so on is well known, but in reality the situation is much more complex, and several different forms can be recognised, particularly ephemerals, annuals, biennials and perennials which may be monocarpic or polycarpic (see below), based on their normal length of life, though some of these are interchangeable according to the conditions.

Ephemerals, as their name suggests, are very short-lived indeed, and they may complete several life cycles within a year, setting seed, dying and germinating again from seed very rapidly. They are an extreme form of the annual plant, whose characteristic is that the life cycle is completed within a year and the plant dies after it reproduces. Normally such plants germinate from seed in the spring, grow rapidly through the warm weather of summer and produce seeds by late summer or autumn, dying afterwards. The seeds then overwinter and germinate the following spring. There are, however, a number of 'winter annuals', particularly common in drier continental areas, whose seeds germinate in autumn and overwinter as seedlings, ready to grow rapidly as soon as favourable conditions return in the spring.

A biennial has a life cycle extending over two years. Normally seed germinates in the spring and the plant grows throughout the first season to produce a non-flowering rosette, usually with a sizeable storage tap root. It overwinters at this stage, and then grows rapidly the following spring and produces flowers and seeds, after which it dies. Foxgloves (*Digitalis purpurea*) are biennials, as are many vegetables like carrots and parsnips in which we make use of the material stored ready for the second year's growth 'spurt'. In fact, biennials may overlap with short-lived perennials that die after flowering, for in unfavourable conditions biennials may take several years to build up enough food reserves to produce flowers, and indeed some habitual annuals may behave as biennials or even perennials in unfavourable conditions.

Perennials live for three years or more. They may die after flowering, and are then known as monocarpic, although they may take several years to reach the flowering stage. Some ground orchids are monocarpic, sometimes taking eight to ten years to reach flowering, as are many bamboos (*Arundinaria* spp.) and Mexican *Agaves*. The great majority of perennials continue to flower year after year once they have matured, though they do not necessarily flower every year; such plants that do not die after flowering are known as polycarpic plants, and they include trees and shrubs, many bulbous plants like tulips or irises, and the majority of other perennial plants.

LIFE SPANS

Ephemerals, annuals and biennials, as we have seen, all have short, fairly well-defined life expectancies. So far, however, we have only defined perennials as living for three years or more, although within this group there is enormous variation in life span. Many herbaceous perennials, such as red clover (*Trifolium pratense*), white clover (*Trifolium repens*), some violets (*Viola* spp.), gentians (*Gentiana* spp.), and many others, have an average life span of two to five years for each plant, which is about what one would expect. However, many small herbaceous plants live for a surprisingly long time; herb Paris (*Paris quadrifolia*) lives for about 20 years, whereas many species of orchids studied over a long period in Sweden lived for a remarkably long time—twayblades (*Listera ovata*) and marsh orchids (*Dactylorhiza* spp.) that had been mature when first studied were still flowering regularly after almost 30 years. The fact that so few of the original population had died leads one to predict that some individuals could live for 70 or even 100 years, perhaps longer than man! Other plants, particularly tussocky ones like some grasses, have a growth pattern that allows them to grow more or less indefinitely even though the centre or base of the tussock may die. Such plants are virtually immortal, at least so far as has been detected up to this time.

Trees and shrubs, as is well known, can live to a considerable age, though again this varies widely between species. Birches (*Betula* spp.) for example, are relatively short lived with a life span of only about 70 to 100 years. Beeches (*Fagus sylvatica*) live for about 200 to 300 years, whereas oaks (*Quercus* spp.) may live for 500 years, or even much more if conditions are right. Some American woody plants live much longer still; for example, individuals of the bristle-cone pine (*Pinus aristata*) are known to be about 5,000 years old, and colonies of the box huckleberry in Pennsylvania are believed to be over 10,000 years old!

Interestingly, it is possible to prolong the life of trees and shrubs considerably by a management technique known as coppicing, and to a lesser extent by pollarding. Both practices have been widespread in Europe for many centuries, probably stretching back to prehistoric times in a few places. Coppicing involves cutting the tree back to its base at regular intervals (seven to twenty years, depending on the produce required) and allowing the shoots to regrow. Pollarding involves the same technique carried out at about 8–10 ft (2.5– 3 m) above ground level to avoid browsing animals. The subsequent regrowth can be harvested endlessly, and the above-ground part of the plant barely ages if regularly cut. There are examples of coppiced ash (*Fraxinus excelsior*) and small-leaved lime (*Tilia cordata*) which have been regularly coppiced and are probably over 1,000 years old. If not coppiced, trees usually eventually become 'stag-headed' as their highest branches die back, and they eventually die either from excessive fungal infection or by toppling in a gale, or both. Coppicing and pollarding prevent this from happening.

Herbaceous perennials, too, do not have a fixed life span for each species. Mowing or cutting back of some species, especially many grasses, seems to prolong their life, especially if it prevents flowering and fruiting, which are such a drain on the plant's resources and may increase the likelihood of dying. In other cases, the environment may affect the length of life; for example, red clover at low altitudes in Russia was found to live for two to three years, whereas at high altitudes it was found to live for ten years, though growing more slowly and flowering less profusely. Other similar examples are known.

In fact, we know very little about how long many plants live, but the general answer to the question is 'surprisingly long'.

THE SEASONAL CYCLE

Most perennial plants of temperate regions have a very definite seasonal cycle of growth, flower production, fruiting and then leaf loss or die-back. What is going on in plants through these changes, and what determines when each event happens?

Throughout the winter, most plants remain completely dormant, either as below-ground storage organs (such as bulbs), 'resting' rosettes, or leafless woody 'skeletons'. In each case, activity is reduced to a minimum but all such plants have food stored from the previous season, either in special storage organs like bulbs and tubers, or spread throughout the tissues. This supply of food allows the plant to respond to a rise in temperature (for most plants the critical temperature is 5° to 6° C) by beginning to grow and put out new leaves up to the stage when this foliage can produce food for itself. From then on, the foliage grows and enlarges, reaching a peak at full leaf, usually in midsummer, though some plants go on producing new leaves through the season.

Flowering time

At some point the plant begins to produce flowers, which may be in early spring, midsummer, or autumn, but at whatever time they occur they tend to be constant for that species. In ephemerals, and some annuals and other species, the time of flowering seems to be determined simply by when there is enough photosynthesis going on to allow materials to be diverted into flower production.

In many other species, the mechanism is much more subtle and is determined largely by the length of the days—a phenomenon known as photoperiodism. Some plants will only flower when the days are longer than a certain critical length (or, more relevantly, when the nights are shorter than a certain length). Spinach, for example, will not flower until exposed for two weeks to days that are 13 to 14 hours long, or more. Such plants are known as long-day plants, and frequently they will not flower at all if moved to the tropics because the days never reach the critical length. Other flowers respond by flowering only when the length of daylight is shorter than a certain critical period, e.g. chrysanthemums and dahlias. Such plants are known as short-day plants.

Knowledge of such reactions is clearly of great advantage to floriculturalists or market gardeners. For instance, chrysanthemums can be prevented from flowering (to keep them for the Christmas market, perhaps) by an additional period of artificial illumination simulating extra day length. They can then be brought into flower simply by discontinuing the extra light at the appropriate time. Intriguingly, it was subsequently found that a very short period of illumination in the middle of the night was sufficient to make the plants not flower: in other words, it is the length of darkness that is the critical factor, and a few minutes of light in the middle of the darkness deceives the plant into believing that the dark period is shorter, so it does not flower. The reverse is true of long-day plants.

In fact, we now know that there are not just the three simple categories of day-neutral, long-day and short-day plants, but a range of about 50 categories needing different interactions of day length and temperature; these are clearly required to prevent, for example, the day length of autumn having the same effect as that in spring. Some plants can respond very quickly to certain day lengths: a plant of the genus *Xanthium*, for example, will be induced to flower by exposure to just one short day of less than $15\frac{1}{2}$ hours, even if subsequently returned to longer days. Equally interesting is the revelation that this provides, namely that plants can somehow measure time and that it can be extremely precise: for example, variations in day length of only 30 minutes are long enough to prevent some plants from flowering, and some tropical plants growing where day length varies only by about one hour throughout the year can detect when it falls below $12\frac{1}{4}$ hours! How do plants measure time, and how do they respond to these changes?

The mechanism of flower production

This is not precisely known, but the general pattern seems to involve a light-sensitive pigment and a flower-inducing hormone. The young mature leaves seem to be the best detectors of day length, and they contain a pigment known as phytochrome. After a period of daylight, this phytochrome exists in one particular energy-rich form known as phytochrome P_{730} (the number reflects the wavelength of light to which it is sensitive). In the absence of light, this gradually reverts to the energy-poor form known as P_{660}, at a constant rate. This seems to be the basic 'timing' method, though it is not quite clear why it affects long-day and short-day plants differently. Somehow, the different relative concentrations of the pigment forms promote or inhibit the synthesis in the leaves of a flower-inducing hormone (sometimes known as florigen) which is translocated out to the developing initials. The substance seems to be the same throughout the plant world as extracts from one 'induced' plant can induce another to flower even in its wrong day length, and similarly the leaf of an induced short-day plant can be grafted onto a long-day plant and encourage it to flower in short days. There are many unanswered questions, but this seems to be the basis of flower timing in day-length sensitive plants.

Other seasonal changes

Once the plant is induced to flower, a train of events is set in motion through pollination and fertilisation to seed and fruit production. Many herbaceous plants will simply continue growing and possibly producing flowers until temperatures become too low for growth and frosts break down the above-ground parts. Woody perennials, however, especially deciduous ones, have a more defined cycle of events, notably that of leaf fall in the autumn. The main difference between deciduous and evergreen trees and shrubs is that evergreens cast their leaves continuously throughout the year, whereas deciduous trees cast them all at once, in the autumn. Why do they cast their leaves at all, and what makes it happen?

Leaf fall seems to serve two functions: it cuts down water evaporation from the plant at a time when transpiration is not necessary (as no growth is taking place) and when little water may be available in the soil (because it is frozen); and it serves as a way of disposing of waste products from the plant

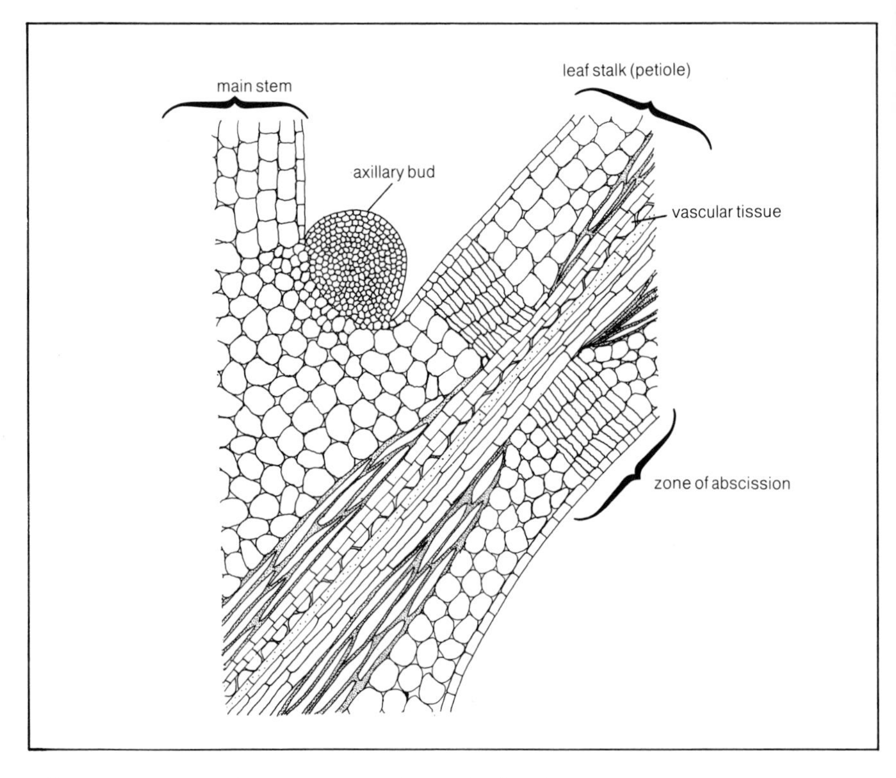

Diagram showing abscission zone at base of leaf stalk.

*This picture of a frost-encrusted holly tree (*Ilex aquifolium*) illustrates the ability of evergreen trees to retain their leaves whatever the conditions.*

body at the end of the growing season. It is not brought about simply by the onset of cold weather, for leaf fall takes place even in mild autumns or frost-free areas, though it may be slightly delayed. Instead, it seems to be triggered by the declining day lengths of autumn, in a similar manner to flower induction.

The leaves do not simply break from the tree—there is a specific mechanism which allows them to fall. There is a layer of cells at the base of the petiole (leaf stalk) known as the abscission layer, visible from early in the growth of the leaf. At the same point, the vascular tissue from the leaf narrows slightly. As autumn approaches, the development of this layer is promoted by hormones issuing from the leaves, and both the ubiquitous auxin (*see* Chapter 3) and another substance (abscisic acid) seem to be involved. It is interesting that parts of trees growing next to street lamps or other light sources may be the last to lose their leaves, as this simulates longer days.

As the leaf becomes senescent, so the abscission layer begins to disintegrate, and at the same time a layer of corky and fatty tissue is formed to protect the exposed surface when the leaf falls. Within the leaf, the green pigment (chlorophyll) breaks down first, exposing the other pigments occurring in low quantities in the leaf, while any useful substances are withdrawn from the leaf. At the same time, over a period, toxic or waste materials are stockpiled in the leaf for removal. The revealed pigments are, of course, the reds and yellows that are such a familiar feature of autumn leaves, especially after a hot, sunny summer when more pigment is manufactured.

Eventually the abscission layer disintegrates completely and the leaf falls. The fall is hastened, of course, by autumn gales and also by frosts which tend to cause the crack created by the disintegrating abscission layer to expand as water enters it and freezes. So, after the loss of all its leaves, the plant becomes dormant with all its life processes slowed to a minimum.

The end of the dormancy period

One final question remains: what causes the plant to break its period of dormancy and begin to grow? As usual, the answer is not simple! It seems that inhibitors collect in the bud, reducing its likelihood of growing, and then a period of cold followed by increasing day length coupled with rising temperatures causes an upsurge in growth-promoting hormones, notably gibberellins, which overcome the growth inhibitors and start growth going. This happens in many plants, but not all, and it is clear that there is a complicated network of internal controls, affected by the external environment, preventing growth before the correct season—and it is clearly highly effective!

A TYPICAL LIFE CYCLE

We have looked at many of the processes that go on in plants in isolation, but not in detail at the life cycle of any one plant. As we have seen, plants are different in many respects, but the following life cycle gives a clearer idea of how at least one species lives its life.

The spring gentian (*Gentiana verna*) is a small herbaceous perennial producing brilliant blue flowers in mountain grasslands throughout the northern hemisphere. The figures given below apply to a population of gentians growing in limestone grassland in the Pennines in England, and plants growing elsewhere would probably behave slightly differently, like the red clover in Russia already mentioned. The seeds are tiny, and are widely distributed by the wind, but studies indicate that only a tiny proportion actually germinate successfully. On germination two small seed leaves are produced above ground, usually in early summer, followed in that year by one or two true leaves. The plant then overwinters in this stage, and the following year the leaves expand into a rosette of at least a dozen leaves and the seed leaves disappear. In its third, or possibly fourth, year the plant may produce a flower. If it does so, it will probably die after flowering, though it need not necessarily do so. Out of a population of 100 rosettes, 50 per cent of those that flowered died before the next season, whereas only 20 per cent of the non-flowerers died. Some may go on to flower in another year, and, extremely rarely, for a third time, but never more than this.

The flower is bright blue and very visible and attractive to insects. Relatively few insects live in exposed mountain areas, but the flower colour and supply of nectar attract those that are around, mainly bumble bees. Pollination success is very high, partly through self-pollination, and the flowers can develop readily into capsules. In grazed situations, though, most flowers never produce mature fruit (because the flowers or fruit are bitten off), but if they do mature there are usually about 160 seeds, with one capsule per plant. Seeds are dispersed, in those few instances where the capsules are allowed to mature, by the wind after the capsule splits and breaks down. Out of the seeds that are produced and dispersed, it appears that very few actually germinate, and the majority of new gentian plants are produced by a purely vegetative method. An established rosette will produce fine stolons which grow underground and emerge up to 10 inches (25 centimetres) away to produce a new rosette, capable of flowering in its second year after emergence.

In general, about 25 per cent of all the rosettes in a population die each year, and the average length of life of a rosette is about three and a half to four years. Some will flower once, a few twice, while a surprising number will live for over five years without flowering, and many simply die without flowering at all. In any given year only about seven plants out of 100 will produce a flower, a much lower figure than one might expect but one that is probably not uncommon in perennials growing in grassland. About one rosette in three will produce a new rosette on a stolon that is successful, but others may be produced that never appear above ground. Seedlings only account for about one new plant in 1,000, so their effect is almost insignificant in numerical terms.

This simplified life history of the spring gentian perhaps gives a better idea of how some of the processes work in real life, and how often many aspects fail: how few plants flower, and even fewer produce seeds; how many seeds go astray, and the number of those that do germinate that fail. Clearly it is a plant that is well adapted to the difficult situation of montane grazed

*The spring gentian (*Gentiana verna*) can reproduce both vegetatively and sexually. This ensures that the species will not die out if the few flowers do not produce seeds, but evolutional adaptation through cross-pollination is also possible.*

grassland in its ability to produce most new plants by underground stolons, while retaining the ability to produce new genetic material from seeds from cross-pollinated flowers. This leads us on to a consideration of the ecology of plants—how they react to their environments and each other, both as individuals and on a world scale.

7 · THE ECOLOGY OF PLANTS

Plants, as we have seen, are not static objects. They grow, behave and react to their environment and to each other. We have also seen that each plant can produce many thousands of seeds in its life, of which only one or two need grow into adults to maintain a stable population. The study of how plants compete and interact with each other and with their environment is known as plant ecology, and although it is perhaps less glamorous than cell biology or genetics it has helped us with answers to many of today's problems as well as posing many more questions.

WORLDWIDE ECOLOGY

In the broadest sense, we can look at the ways that plants have grouped themselves according to the great climatic regions of the world. The broad characteristics of the vegetation are determined largely by climate as affected by latitude, altitude and more local factors, and these are described below.

1) Forests. Forests of the world fall broadly into the four categories of coniferous forests (softwoods), temperate hardwoods, tropical hardwoods and mixed coniferous and hardwood forests. Overall, about 20 per cent of the land surface of the world is forested.

   Coniferous forests account for about 35 per cent of the world's forests, and are mainly in the northern hemisphere, with extensive spruce, pine and fir forests across northern North America, Scandinavia and Russia, and outliers further south to the Mediterranean, down the Rockies and throughout the Himalayas, generally following the cooler zones.

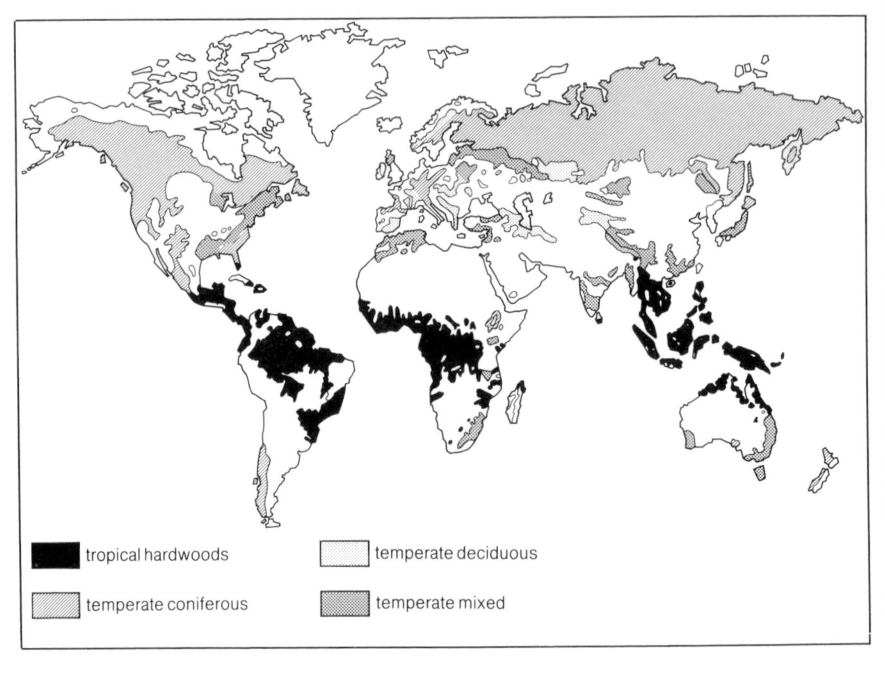

Map of world forest regions.

Temperate hardwood forests occur throughout the temperate areas where there are significant seasonal changes, and form a substantial part of the forests in central Europe and the USA, parts of Asia, especially China and Japan, and in New Zealand, Australia and South America. They reach their greatest dominance in areas such as Britain and New Zealand, in the middle of their range.

Tropical hardwood forests can be divided into tropical rain forest and tropical deciduous forest. Tropical rain forest is probably the lushest and most varied vegetation type in the world, growing in areas where there is virtually no seasonal change at all. Conditions are perfect for plant growth, with continuous high temperatures, no drought and good light all the year round, so that plants of tropical rain forests are almost invariably evergreen since there is no need for a dormant season. The number of species involved is enormous as they all jostle together in perfect growing conditions; over 400 species of tree alone have been recorded in 3 square miles (7.8 square kilometres) of jungle in Brazil, and the number of shrubs, herbs, ferns, mosses and so on below them must be many times that number. The forests are also characterised by the enormous numbers of epiphytes (plants growing upon other plants, using them simply as a surface) because of the high rainfall and humidity. Regrettably, tropical rain forests are now being destroyed at an enormous rate and their richness and diversity is rapidly disappearing.

Tropical deciduous forest occurs in areas of the tropics where there is a seasonal change, usually between wet and dry seasons, such as the monsoon belts of south and east Asia.

Mixed coniferous and hardwood forests may occur wherever conifers occur, though usually they form transition zones between two types.

2) Grasslands. Two widespread types of grassland are identified as occurring on a world scale—savannah and steppe grasslands.

Savannah grasslands generally occur in dry tropical or subtropical conditions, and usually contain isolated trees. They are the *llanos* of Venezuela, the *campine* of the Congo area and the *Patanes* of Sri Lanka. *Steppe grasslands* occur in dry temperate zones, and they include the *prairies* of the North American corn belt, the *steppes* of south Russia including the Ukraine (perhaps the biggest area of grassland in the world) and the *pampas* of South America.

Such classifications of broad vegetation types are, of course, gross generalisations useful only on a world scale. Another way of looking at world plant life is to classify it into biogeographical areas, based on the species involved and their presumed centres of evolution and movement. Such regions are usually more or less isolated from each other by some major barrier such as a mountain range or an ocean. Six broad divisions, with subdivisions, have been recognised, and they are known as the Boreal (roughly the northern temperate regions); the Palaeotropical (divided into African, Indo-Malaysian and Polynesian); the Neotropical (most of South America); South Africa; Australia (including New Zealand); and Antarctic. Opinions, however, differ on the exact divisions, but it is one way of subdividing the world's plants.

Overleaf: The flora of hay meadows, which are managed without fertilisers or weedkillers, is extremely varied. There are at least eight different species of herb in this example.

Inset: Heathland embodies small-scale variations in vegetation which may not immediately be apparent. In this photograph, for example, green patches of bracken can be seen on the far slopes. These occur in hollows where the soil is much richer.

COMMUNITIES AND POPULATIONS

The conditions under which plants grow, both throughout the world and more locally, are endlessly variable. Temperature, light, day length, humidity, rainfall and period of snow cover are some of the more obvious factors, but there are also many other less obvious factors such as aspect and shade, the level of nutrients in the soil and even the relative proportions of different nutrients. Perhaps the most interesting factor, yet the most difficult to quantify, is the interaction between different plants. Throughout the world there are many hundreds of thousands of species of plants, yet, under the pressure of competition and natural selection, they have all evolved slightly different requirements and they exploit their environment in a slightly different way. Most plants can only be successful in a very narrow range of circumstances—within a certain temperature range, with a certain light level, and under certain conditions of nutrient status in the soil, for example—and away from their optimum conditions they lose in the competition with other species.

People often ask why it is that, for example, if a gentian from the Himalayas can survive in an English garden it does not occur here naturally? Apart from the problems of distance involved in this instance, it has to be remembered that garden plants are peculiarly cosseted, and we are only seeing one part of their life cycle. In the wild they have to attract pollinators, produce adequate fertile seed, and disperse it to suitable places; this seed has to germinate and then compete successfully with all the other seedlings growing around it, or (more likely) with already established vegetation. Then some, at least, have to survive to maturity to produce flowers and fruits again. Most important of all, perhaps, it has to do this every year so that the species can continue to survive in that place. Garden plants are absolved from the necessity to do all these things, as we weed around them, harvest the seed before it rots, sow the seeds in boxes of compost instead of letting them take their chance, perhaps cover them over in wet winters to prevent them rotting, and then if they die off we simply replace them! So, plants can survive in environments different from that which they normally occupy, but they are most often prevented from doing so by competition from other plants at all stages, and they can only win the battle in the conditions that suit them best.

The countryside is not, however, just a random collection of plants all jostling together to survive. We can recognise distinct vegetation types that have particular recognisable features. In most parts of the world the vegetation has been affected by man for so long that the boundaries are no longer natural ones. An oak wood, which is one vegetation type, will rarely extend as far as the environment lets it; it will be constrained by arable fields, pasture or other man-induced vegetation features or developments. In less affected areas, such as high mountains, there are still distinctive vegetation types, with the boundaries between them determined by, for example, increasing altitude, a change in soil type, or the edge of a flood plain.

Such recognisable types occur because, within certain limits of the environment, one species or a group of species can become dominant in that situation. These are most frequently tree species, since their size allows them to dominate other plants most successfully. For example, on the southern slopes of the Himalayas, silver fir forest is dominant over wide areas of countryside at an altitude of 10,000 to 11,500 feet (3,000 to 3,500 metres). Above this height, the colder winters and greater snowfall do not suit it, and junipers and birches become dominant in its place; lower down than this, trees that cannot stand the cold but are more aggressive in warmer conditions, such as oaks of various species, tend to replace it. At the same time, the dominance by one or more species provides conditions that are suitable for a wide range of other species that like the same soil and general conditions and can tolerate the shade or partial shade of the woodland. Generally speaking, the plants of the silver fir wood will be different from those of the adjacent oak wood, though a few species will tolerate both. Often the former are species that are specifically confined to a particular vegetation type or environment, and they may be taken as indicators of that environment.

Even within a vegetation type that is broadly similar and dominated by one species, there are frequently many small-scale variations that are reflected in the vegetation. In much of Northern Europe, for example, and especially in Holland, France and Britain, the predominant vegetation on acidic and free-draining soils is heathland. To the untrained eye, this appears to be exceedingly uniform vegetation, though in fact it is far from being so. The predominant plant is usually ling (*Calluna vulgaris*); in the drier areas, however, it will often be co-dominant with another heather, the bell heather (*Erica cinerea*). In parts where the soil is more clayey, or where the drainage is poor, there will be a higher proportion of the cross-leaved heath (*Erica tetralix*) while the bell heather disappears under such conditions. There may be patches of common gorse (*Ulex europaeus*) which almost certainly will be growing on the sites of old fires, or where the soil has been disturbed (they may be so persistent that they can be taken as indicators of archaeological sites in some cases). There will also be areas that are dominated by bracken (*Pteridium aquilinum*) and these are usually areas with a deeper, more nutrient-rich soil, perhaps in a hollow that collects the nutrients from the surrounding area. Elsewhere, there may be patches of bare soil that support different plants like the mossy stonecrop (*Crassula tillaea*), and here and there may be small grassy areas on better soil which are perpetuated by grazing animals. Thus, even a relatively uniform habitat like heathland is full of variety and something like an oak/ash wood is infinitely more so, with its endless variety of 'niches' for plants to colonise.

SUCCESSION

Vegetation is in a continual state of change, both for natural reasons and artificial ones. If a piece of ground becomes bared for some reason—such as a fire or a clearance by man for cultivation—it will not remain as a clear area for long (unless managed as such), but will revert within 100 years or so (considerably less in the tropics) to a woodland. The process of changing from bare ground to woodland is known as succession. It is of particular interest because of the way in which each group of species that colonises tends to change the character of the area slightly so that the next species in the series are better able to invade.

In a natural succession from, say, bare rock exposed by a retreating glacier or a newly-created volcanic island, the earliest colonisers are frequently lichens and mosses on rock or very shallow pockets of soil. The lichens, in particular, are creators of conditions suitable for higher plants to colonise since they secrete acids and push their filaments between the rock particles, laying them more open to erosion. Gradually, increasing amounts of soil form and more nutrients are released, so that flowering plants, and especially annual plants, can colonise. In a secondary succession, from a man-made clearance, these early stages are short-circuited as the soil already exists. The annual weeds will be rapidly followed by more persistent perennial grasses and then shrubs such as hawthorn (*Crataegus monogyna*) and blackthorn (*Prunus spinosa*). These are usually followed by the coloniser trees—the weeds of the tree world—such as birch and pine. Birch in particular colonises readily, but is relatively short-lived (60 to 70 years) and its place is gradually taken by species such as oak, which is slower growing but much more persistent.

This series of stages, which varies in the species involved according to the part of the world and the local conditions, is known as succession. All such series share the characteristics of increasing amounts and complexity of vegetation as they progress, and increasing numbers of animal species dependent on the vegetation. In virtually all circumstances, except high mountains, towards the poles, and some other extreme conditions, the

Rosebay willowherb (Epilobium angustifolium *) has an 'opportunist' strategy, rapidly colonising clearings or burnt areas by means of its wide-ranging wind-borne seeds.*

vegetation at the end of the series is woodland, and this stage is more stable than any of the preceding stages.

CLIMAX VEGETATION

This leads to the idea that, in any given area, there is a stable vegetation that was there before man altered it, and which will return when man's influence ceases; in the case of the abandoned arable field in a temperate area, as described above, the final stable vegetation would be oak woodland. This stable, ultimate vegetation is known as the climax vegetation. Unfortunately, the matter is not so simple, for several reasons: firstly, the climax vegetation in any one place depends, amongst other things, on the climate, and the climate is always changing. For example, in parts of North America the climax vegetation of 11,000 years ago was spruce and fir. This was succeeded as climax by oak and pine, then beech and hemlock, then oak and hickory, until today the climax consists of hemlock, chestnut, maple and pine. The same type of changes involving different species have taken place in Northern Europe as the climate warmed considerably after the last ice age and then cooled again.

The second problem is that there are few areas which man has not affected. In many cases this is obvious, and no-one would mistake a field of barley for a true climax vegetation. In other cases, however, the changes are much less obvious, as illustrated in the following two examples. Heathland has been part of the British scene for as long as documentary evidence goes back, and there are many literary and historical references to it. It might be assumed, therefore, that heathland is a climax vegetation on some poor, acidic, soils. We now know, though, that heathland is an artificial environment created and maintained by man. The areas were covered by forest which was cleared by Neolithic or Bronze Age man for grazing or cultivation. In the absence of modern fertilisers, the cleared soils became progressively more infertile and acid so that the vegetation most suited to them was heather. At the same time, the grazing of wild and domestic herbivores prevented trees from re-establishing, and frequent fires compounded the effect. In fact, if grazing and burning are stopped, most heathland soils revert rapidly, through a birch and pine stage, to some form of mature woodland, usually with oak, so heathlands are not a true climax at all, but an artificially-maintained and created situation (known as a plagioclimax).

The second example is less straightforward. It has always been assumed that in Britain and parts of Northern Europe the climax woodland on good soil is pure oak woodland since those fragments of unplanted woodland that do remain are usually oak-dominated. It has been gradually appreciated, however, that this too is something of an artefact of man's activities since oak has been actively selected for over the last 300 to 400 years, and other trees discriminated against. The true present climax would probably contain considerable quantities of ash, lime, beech, elms, maple and other trees, depending on the soil conditions, and oak would be just one of the features of the wood.

So, the concept is not quite so straightforward, but it is a useful idea if not taken too far.

THE STRUCTURE OF VEGETATION COMMUNITIES

Most natural or nearly natural stands of vegetation have a very complex structure which involves hundreds of species of plants and thousands of species of animals. In any one area, no two species occupy exactly the same 'niche': in other words, all species in a community are exploiting slightly different aspects of the environment—some may grow in the gaps where trees have fallen, some in deep shade, and others on the trees themselves; some will choose the poorly-drained parts, others the drier parts, and so on. The divisions are even finer than this, however, since even plants that appear to grow in similar places are actually ecologically separated. The moss and lichen community growing on an old oak tree may consist of tens, or even a hundred or more, of species each growing where the combination of light, nutrients, humidity, and rainwash suit it best.

In a woodland, for example, there is a very wide range of 'niches', each offering slightly different conditions to whatever plant can exploit it. In general, therefore, the longer a woodland type has existed as a type, and the longer it has been in a particular place, the more the 'niches' will have been filled, as both distribution and evolution take time. Oak-based woodland is, as we have seen, the natural vegetation of much of Central to Northern Europe, and the vegetation type has existed for a long time. An old oak woodland, therefore, and particularly one that has survived directly from the original woodland cover following the last ice age, will be very rich in species.

Unfortunately, such woodland is rarely unmodified by man. Where it is more or less unmodified, it has a definite structure, known as high forest (mimicked by many man-made systems of woodland management), in which there is a fairly dense canopy of mature trees of varying ages, with a distinct layer of shrubs and saplings underneath it, and a herb layer close to the ground. In its natural state, oak is a very variable tree, and individuals branch in different ways, have different growth rates, and live for different lengths of time. This means that the trees never all die at once, but instead they die at irregular intervals, creating gaps in the canopy that are then colonised by young trees of oak or another species. Elsewhere, within the canopy, there may be trees of ash, lime, beech, elm and many other species, and each of these will cast a different canopy density, produce a different leaf litter, and come into leaf in spring at a different time.

Hence conditions under the canopy are never uniform, and a range of shrubs is able to survive below the canopy and sub-canopy trees. These, in turn, cast different densities of shade and produce different litter—for example, the conditions under holly (*Ilex aquifolium*) are very different from under hazel (*Corylus avellana*), and very few species grow under the former.

Plants growing at or near ground level are, inevitably, highly affected by the trees and shrubs growing above them, and many species of herbs are characteristically associated with one type of woodland that suits them best. In the characteristic, natural, type of woodland that has been described, the variety in type and density of the canopy leaves opportunity for many plants to survive: the well-adapted woodland plants that flower in spring before the leaves on the trees are out, and the less well-adapted species that flower in summer and need more light. At the same time, the variety in forms of tree, together with the high humidity and variation in light intensity, allows an enormous range of epiphytes (plants growing on other plants without harming them in any way) such as ferns, lichens and mosses to survive on the trees.

Every layer of the woodland below the trees is used to the full, and this means that an enormous amount of vegetation is produced in a relatively small area, as well as a wide range of species. Naturally, this means that a vast range of plant-eating insect life can survive, rich both in numbers and in species because of the amount and variety of food sources available to them. As long as there are suitable nesting and breeding sites available—and in a natural varied wood there usually will be—this plethora of insect life allows a wide range of insectivorous birds such as tits, warblers and others to become part of the community, and of course the variety of fruits (hips, haws, hazelnuts and so on) allows a range of fruit-eating or omnivorous birds and mammals to survive. This wealth of 'meat' allows the higher predators, such as sparrowhawks, to do well in the presence of adequate food supplies. In other words, in the same way as plants are central to life on earth, so they are central to the working of an individual community such as a deciduous woodland.

COPPICE WOODLANDS AND PLANTATIONS

Throughout the Middle Ages, and probably well before, trees and shrubs were recognised as a source of fuel, building material, fencing material, fodder and many other things. Very early on it was discovered that a tree, once cut or even burnt, will simply regrow from the base into a series of smaller trunks. Each of these stems can be cut when required and the process repeated, apparently almost *ad infinitum* if required. This process of cutting back trunks to form sprouting shoots at ground level is known as coppicing, and it was the dominant form of woodland management up until

In a coppice such as this, where the hazel trees have just been cut for hurdle-making, leaving the stumps visible in the background, the newly increased amount of light enables bugle, solomon's seal and primroses to flower superbly.

the present plantation era. This gave rise to a rather different woodland structure in which the primary component was coppiced trees and shrubs (latterly mainly hazel, but many other species were involved), but with a sparse layer of canopy trees, perhaps 15 to the acre (37 per hectare), spread through the wood. This system was known as coppice-with-standards woodland.

Although poorer in total mass of vegetation and probably poorer in species numbers than the natural high forest, coppicing had the great merit of providing a rotational cycle of light and shade which allowed all the light-demanding flowers to bloom in profusion every few years and then retreat into dormancy. It copied the varied conditions of the natural forest in a more organised way, and many species flourished on it. The process was so widespread that much of the wildlife and plant-life of such areas has been reduced to species that favour this particular management.

Regrettably, this type of management has now virtually ceased, and we are left without either natural woodland or managed coppice. It has been replaced instead by plantation forestry, as timber products have become more valuable than those of the underwood. This normally involves clearing the existing vegetation and replacing it with trees planted in rows—normally of species alien to the area—which are then kept free of intrusive weed growth such as birches. The density of planting and the acidity of the leaf litter usually ensure that nothing survives under the trees. Most plantations, therefore, consist of one dominant tree with a very few plant species on the ground and a few insects living on them, in stark contrast to the incredible three-dimensional mosaic of diversity that comprises natural woodland.

THE ECOLOGY OF SPECIES

We have looked, so far, at the way in which all plants react together with each other and with their environment to produce definite communities and vegetation types, both on a world scale and more locally. An alternative branch of plant ecology involves looking at individual species and the ways in which they survive and reproduce, known as autecology. It is immediately obvious that some plants are much more common and more successful than others in any given place—but why should this be so?

The process of finding out more about individual plants takes place on two levels: extensive survey to determine the distribution and requirements of the plant, and intensive research to ascertain more about what is happening in the populations of the plant—how long individuals live, what makes them die, what eats them or infests them, how pollination takes place, how much seed is set, and so on. In many countries of Europe and North America, the former investigations have been going on for hundreds of years as botanists have amassed data on the distribution of plants and the type of habitat in which they grow best. For a long time this data was summarised, if at all, only in a very general fashion such as 'widespread but local in woods', but for the last 20 years or so the process has become much more precise. Many recording schemes are in operation, usually using information from amateur botanists, that collect and analyse all records received, and, with the aid of computers, produce maps to show the countrywide, national or even international distribution of species. This immediately gives a much clearer indication of what the distribution of the species is likely to be related to, and also indicates what species are rare or becoming rare.

As this sort of information is now more readily available—though it is never a complete record—attention has turned to finding out more in detail about plants. For many species, even in a populous and well-studied country like Britain, we have little idea of how long they live, how many new plants are coming into the population, or even how they are pollinated, let alone the answers to the more complex questions of their relationship to their environment and other species. There is a fruitful field for research awaiting anyone prepared to look at one species over a reasonable time period to find out exactly what it is doing.

8 · PLANTS AND MAN

We have seen throughout this book that plants are central to the existence of life on earth. This is something of which most people are only vaguely aware, if at all. Yet there are many more specific ways in which we use and relate to plants—as food plants, in our gardens and houses, and as sources of medicine and materials.

PLANTS AS FOOD

Man has been eating plants ever since he first evolved as a recognisable species, but for the greater part of that time plants were simply used as and when they were available, in the same way that wild primates use food sources. For the last 10,000 years or so, however, man has learnt to cultivate plants so that a reliable food source is available when and where it is required. Before this, there would never have been any consideration of improving the food source genetically; if a food source was poor, something else would have been tried. Nevertheless, as soon as domestic cropping was practised, with the collection of seeds for next year's crop, it is highly probable that a form of selection of the best strains began; seeds from the best plants were saved for the following year, any chance worthwhile mutations or hybridisations with wild plants would have been retained, and the most disease-resistant plants would automatically have been kept.

The beginnings of settled cultivation, which seems to have arisen separately in different parts of the world, gave rise to the great early civilisations as food supplies became stable and provided surpluses. All of these early civilisations depended on some form of cereal cultivation, notably wheat in the middle eastern and southern European cultures, rice in southern Asia, and maize in South America. By about 5,000 years ago, a very wide range of crops was being grown, and it is remarkable that virtually no new major staple crops have been introduced since that time.

In many cases, gradual selection of the crops most suited to human use (rather than for survival in the wild) has considerably altered the crop plants from their original appearance and characteristics. Bananas in the wild, for example, produce small, rather inedible, seedy fruit in contrast to the large seedless fruit grown for human consumption; and the ears of wild wheat are rough and stiff with the grains tightly enclosed for protection.

Other plants have undergone so much change, through selection and possibly chance hybridisation, that their wild ancestors are no longer recognised today. Maize, for example, appears to have derived from one small-eared species which was grown at first as a rather poor crop. At some stage this hybridised with another wild species to produce a much better plant called teosinte, more akin to modern maize, and grown by the South American Indians at the time of Colombus. Since then further crossings between teosinte and the original species have produced the large-eared varieties grown today throughout the world. Similarly, cucumbers and pumpkins have no wild ancestors that can be established with a high degree of certainty.

Other plants, though remaining broadly similar to their wild ancestors, simply proliferated into an enormous variety of forms to suit all palates and growing conditions; all the numerous and varied apples of the world developed from the wild crab apple, for instance, by casual selection and, later, by planned breeding. So, the greater part of the selection of our present crop plants came about entirely by casual selection, chance mutations and hybridisation, with no understanding of the basis for the changes.

It was not until 1866 that Gregor Mendel, an Austrian monk, discovered the basis of inheritance and laid the foundations of modern genetics and plant breeding. He found by a series of carefully controlled and recorded experiments that it was possible to predict the outcome of any 'cross' between two plants with great accuracy, and that inheritance was in fact governed by a series of simple rules, which he outlined. In fact, Mendel's work was only appreciated by about 1900, and plant breeding using these principles really only got under way from the 1920s.

From then on it has become possible to select desirable characteristics and breed them into varieties as required. At first, although the principles were understood, there were many barriers to success; for example, it was frequently found that two plants, whose characters it was desired to mix, could not actually be persuaded to breed together. Gradually, ways of overcoming these barriers have been found, and a major leap forward was made with the discovery and use of a substance called colchicine (isolated from the autumn crocus, *Colchicum*). In cases where the chromosomes of the two plants to be crossed are very different in number or size, the union is frequently sterile because the reduction division to produce the male and female cells cannot take place since the chromosomes cannot pair up. Colchicine tends to promote a spontaneous doubling of the chromosome number, which then gives each chromosome a ready-made partner, and reduction division can take place, allowing offspring to be produced. Breeding for new food plants is a long business requiring many seasons before the work can be completed, though the use of rapid growth environments, and manipulated day length, encourages plants to mature and breed earlier, shortening the generation time. At times, new 'sports' or mutations arise by chance and, if they have characters of value, these can be incorporated into a breeding programme whether they arose in an amateur's garden or on an experimental research station. Red and green cabbages, cauliflower, broccoli and sprouts were all spontaneous mutations from the original wild cabbage.

The 'green revolution'

The last 40 years have been characterised by attempts to increase the world's production of basic staple crops by breeding high-yielding nutritious varieties. In 1943 an ambitious programme was embarked on in North America to try to improve the strains of wheat then available, particularly for use in poor, arid areas like Mexico. The first real fruits did not emerge until 1961 when a dwarf high-yielding yet very tolerant variety was produced, and within a few years this had caused yields in Mexico, and later worldwide, to rise dramatically. The success of this stimulated new programmes to improve rice, set up in 1962, and later maize. Yields of rice were soon increased by the development of a high-yield strain known as IR–8, and maize, which was high-yielding anyway, was greatly improved by the incorporation of much higher protein levels in the grain. One other success should be mentioned, namely the recent production of a new and highly promising crop called Triticale from a hybrid of wheat and rye. These dramatic successes, and many others, have become known as 'the green revolution', as they have revolutionised world food production.

More recently, a good deal of attention has been turned to finding cheap high quality sources of protein. The use of meat is a highly wasteful way of producing protein, as so much is lost in the conversion of grass, grain or other plant material into meat protein; soya beans yield some 10 times as much protein per unit area as chickens, and 13 times as much as beef cattle.

Amongst the alternatives studied have been algae, including blue-green algae, bacteria, and yeasts and other fungi, and many of these are proving very promising though no really satisfactory breakthrough has yet been achieved. Surprisingly little attention has been paid to improving the numerous varieties of beans and pulses available in cultivation or in the wild, or to making them fit for growing in cooler climates.

GARDEN PLANTS

Although the main thrust of breeding has been towards the improvement of food plants, garden plants have been far from neglected. The Chinese and Japanese are known to have been breeding and improving garden plants for many centuries and some of our garden plants are the results of their efforts. Until recently though, most plants in western gardens were collected directly from the wild all over the world, by plant collectors, missionaries and others with an interest in plants. As long as collected plants were able to breed successfully in their new environments, the work of selection immediately began. Much of the work of improving garden plants has fallen to amateurs, either by gradual improvement of strains over a long period, or by the selection and cultivation of unusual varieties (often caused by mutation). Innumerable such 'cultivars' exist, many of them named after their discoverers, and they frequently look nothing like the species from which they have originated. It is mainly the widespread commercial species such as roses, carnations and chrysanthemums that have received the attention of plant breeders, and many new varieties of these have been produced by deliberate breeding and selection.

MATERIALS FROM PLANTS

Not only do we eat plants in innumerable guises, but we also make use of them in many other ways. Wood is perhaps the best-known and most widely-used of plant materials. There is evidence that wood was used in the very early stages of civilisation, and it has been continuously used ever since for building construction, vehicle manufacture, tools, ships and boats, furniture and other items, as well as its ubiquitous use as fuel. The different properties of different species were recognised very early, and there is evidence to suggest that even Bronze Age man in Europe managed trees to provide wood in the form that he required. Throughout the Middle Ages, wood was of paramount importance for house and barn construction, and in ship-building. Many buildings and artefacts of the period were so well constructed, and the timber so well chosen, that they still survive today. The remarkable fact about wood is that, despite the enormous number of competing artificial products such as plastics and metals, wood still remains the most desirable and attractive of all materials, and demand is as high as ever.

Although paper-making probably originated in China in about AD 105, it was not until 1800 that wood pulp was first used in paper manufacture. Since then, almost all paper has been made from wood pulp, and enormous quantities of trees, mainly from Canada, the USA and Scandinavia, are felled each year to make paper, and especially newsprint. Most of the trees used are faster-growing conifers and a few broadleaves such as aspen and birch, and much now comes from plantations.

Other materials derived from trees include cork from the cork oak (*Quercus suber*). This is mainly grown in the Iberian peninsula, and the product is used for stoppers, floats, insulation and other uses. The cork itself comes from the bark of the oak.

Various useful substances are extracted from trees, most notably the tannins used in the tanning industry for converting hides to leather. In England, the practice of growing and coppicing oaks was widespread until about 1900 and the bark was used to provide the source of tannin. Oaks are now rarely used, and other trees provide the tannins.

Rubber is made from latex, a milky-white fluid produced by many plants including such herbaceous plants as the spurges and dandelions. The main commercial source, however, is from the latex of the rubber tree (*Hevea brasiliensis*), a native of South America now widely grown in plantations there and in South-east Asia. Although about half of the world's rubber supply is manufactured synthetically without the rubber-tree extract, several million tons of natural rubber are still produced each year.

Most oils that form the basis for all sorts of industries come from plants. The seeds and fruits of numerous species of all different types are rich in natural oils, thanks to the provisioning of the seed endosperm by its parent. Oils are mainly used in margarines, cooking oils or salad dressings if they are edible, or in industry as paint and varnish bases or lubricants. The main sources of edible oils are soya beans (soya oil forms the basis of many cooking oils and margarines), sunflower seeds, maize (corn oil), peanuts, safflowers and olives, depending on the part of the world and the taste and qualities required. Some, such as sunflower oil, are higher in polyunsaturated fats and lower in cholesterol and are thus gaining in popularity in health-conscious diets. Most of the oil-producing crops are grown in or near the tropics where oil production tends to be highest, though maize and sunflowers, and more recently oilseed rape, can be grown much further north. The oil is extracted by pressing the material in large presses: the residue from which the oil has been extracted is used for making cattle-cake and other feedstuffs, as fertiliser, or ground into low-fat flour.

Inedible oils include linseed oil, from the seeds of flax (*Linum*) and tung oil from the tung plant (*Aleurites fordii*). Both have the property of drying quickly to a thin stretchy film, which provides the basis for paints, varnishes, polishes, lacquers and other widely-used substances.

A relatively recent discovery is the oil from the Mexican jojoba plant (*Simmondsia chinensis*). This has been found to have very similar properties to the oil of the sperm whale, and jojoba oil is at last replacing whale oil in cosmetics, soaps and similar compounds. Its commercial use provides a realistic alternative and some increased prospect of saving the whales from extinction.

The resin from some pines is used to produce turpentine, widely used in industry as a solvent and base for paints and varnishes. The residue after turpentine is extracted is known as rosin, used widely in adhesives and paints.

Many fibres in use today come from plants, and despite the advent of numerous artificial fibres they are still widely cultivated and used. Cotton is probably the most important. It derives from the fluffy white hairs on the seeds of the cotton plant (*Gossypium*), a member of the mallow family. The hairs are removed by a process called ginning, and then cleaned, carded and spun. Cotton has been used as a fibre since at least 3,000 BC in India. Hemp, whose fibres are used to make rope and twine, has been cultivated for even longer, since 4,500 BC in China. It is the oldest known cultivated fibre plant.

The stems of flax (*Linum usitatissimum*) are used to produce the material linen. The stems of the flax are soaked in water until the pectin linkages begin to rot, allowing the fibres to separate; they are then removed from the remaining plant material and dried and spun. The process is known as retting, and has been known since as far back as 3,000 BC in Egypt, where it probably originated. Jute and sisal are two other fibres, from the plants of *Corchorus capsularis* and *Agave sisalana* respectively.

MEDICINES, SPICES AND PERFUMES

Plants contain many biologically active substances which may act as poisons, medicines or stimulants, and a proportion of these are extracted and used in medicines and drinks. Although the majority of drugs and antibiotics are now made synthetically, most of them were originally discovered in some form or other in plants.

Aspirin, now readily manufactured synthetically, was originally discovered as a related substance in the bark of willow trees (*Salix* spp.). Cocaine comes from an extract of the coca plant (*Erythroxylon coca*), a native of Peru and Bolivia, and opium and morphine come from the seeds of the opium poppy (*Papaver somniferum*). Quinine, a traditional cure for malaria, comes from the back of the yellow-bark tree (or cinchona), a native of high altitudes in the northern Andes. The effects of many of these substances were well known to the local people long before modern chemical synthesis, and the active ingredients were extracted by infusing or simply chewing.

The active ingredient of many contraceptives comes from a South American plant, and penicillin, the first antibiotic, was isolated from the colonies of a fungus (*Penicillium notatum*) after it was noticed that bacterial colonies adjacent to it died. Many poisonous plants produce extracts which are useful in medicine in small amounts, e.g. digitalin is a heart stimulant isolated from foxgloves (*Digitalis purpurea*), and atropine and belladonna are isolated from the deadly nightshade (*Atropa belladonna*).

In several parts of the world, particularly the Andes and the Himalayas, many species of plant are collected by the local people and sold or used as highly-valued medicinal remedies. Little is known of the active ingredients, but it is likely that many of the compounds will be active medicinally. The majority of plants have not been examined for medical activity, but there is now a concerted effort worldwide to examine the effects of as many species as possible. There is some urgency since so many species face extinction as their habitats are lost or changed irrevocably.

Many plants, especially those from certain families such as the mint family (Labiatae) and the blue gum trees (*Eucalyptus* spp.), produce their own volatile compounds known as essential oils. These find a wide range of uses, such as oil of peppermint from the mint (*Mentha piperita*) which is widely used for flavouring or scenting sweets, toothpastes and mouthwashes. Oil of camphor from the camphor tree (*Cinnamomum camphora*) is used in various medicines and local anaesthetics, and in the manufacture of celluloid, and eucalyptus oil is widely used as a nasal decongestant.

Herbs and spices all come from plants, not surprisingly. Many plants of warm Mediterranean areas have pleasantly aromatic edible foliage, and these are widely used as pot herbs: thyme, rosemary, basil, marjoram, oregano and numerous others are all members of the mint family, already noted as being a particularly aromatic group. Spices, in contrast, more frequently come from woody plants, usually growing in the tropics. Cinnamon comes from the bark of the cinnamon tree (*Cinnamomum zeylanicum*), a native of Sri Lanka (Ceylon), which is still the largest producer. Pepper comes from peppercorns which are the dried fruits of a perennial climber, *Piper nigrum*, native to India and widely grown there and elsewhere in the tropics. World-wide trade reaches some 75,000 tons per year. Vanilla is extracted from the enormous capsules of a climbing orchid, *Vanilla planifolia*, native to Central America, though now grown elsewhere in areas of similar climate. Nutmeg and mace are both the products of the same tree, *Myristica fragrans*, a native of the Moluccan Islands. Nutmeg is the seed whereas mace comes from the fleshy part of the fruit. The list of spices is endless, but, in general, most come originally from oriental tropical regions, especially India and China, and most come from trees or shrubs.

Perfumes, too, are usually plant products. Scented flowers such as roses, hyacinths and acacias, or the foliage of lavender, geraniums and rosemary, provide most of them, though some are now made artificially, once the structure of the active ingredient has been established.

Finally, many of the great drinks of the world rely on plants. Tea is made, of course, from the leaves of the tea plant (*Camellia sinensis*), a native of south-east Asia but now widely grown in the subtropics and warmer temperate regions. Coffee comes from the ground beans of the 'arabica' coffee plant (*Coffea arabica*), or the 'robusta' coffee plant (*Coffea canephora*). Both are now grown worldwide with enormous exports, and numerous different varieties exist. Cocoa comes from the dried ground beans of the cocoa tree (*Theobroma cacao*) which originated in the American tropics (though it is now largely grown in West Africa), and the popular tropical drink maté comes from the leaves of a relative of holly, *Ilex paraguensis*, grown mainly in South America.

Most alcoholic drinks derive their alcohol, and frequently their taste, from plant products. The most common sources of fermenting material include barley, rice, potatoes, molasses and grapes, and drinks are often flavoured with some other plant material such as hops in beer, juniper berries in gin, or aniseed in several drinks.

ALL-PURPOSE PLANTS

Most plants are used for one particular product. A few trees, however, occupy a central place in the local economy by producing a wide range of products. Coconut palms (*Cocos nucifera*) and date palms (*Phoenix dactylifera*) are particularly good examples. The date palm, for instance, provides material for roofing, basket-weaving, cloth, furniture, saddles and fuel from its leaves; in addition the fruits are a staple nourishing food and a source of cash, being made into various drinks and vinegars, and occasionally used as

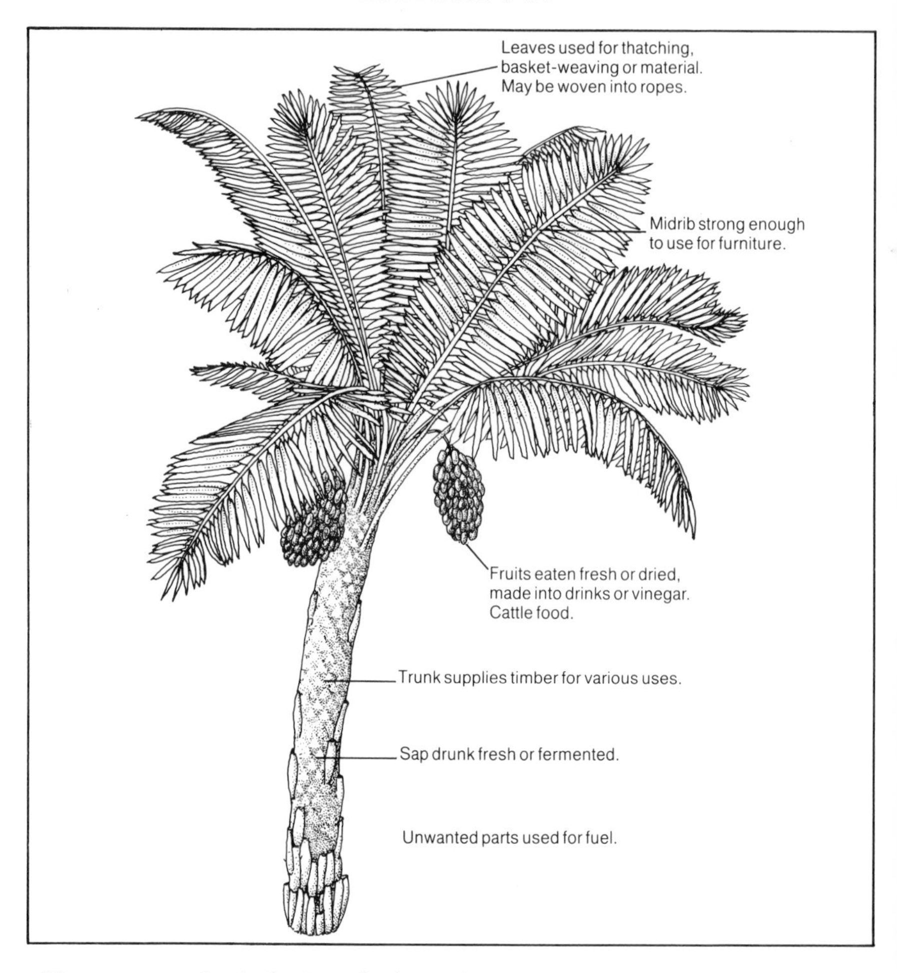

The many uses of a single plant, the date palm.

an animal foodstuff. The trunk supplies timber for building houses or huts, and is used for all parts of the work including window frames, doors and rafters. It can also, of course, be used as fuel. Finally, the sap is used as a refreshing drink, or fermented to make an intoxicating drink. What more does a community need?

CONSERVATION AND EXTINCTION

Man has always affected his environment. At first we were little different from any other wild animal and the effects were quickly healed. As soon as the first settlements began, the effects became more marked and areas were destined never to return to their original vegetation. As the population of the world grew so did its effects, but it was not until the industrial revolution that our effect on the environment really accelerated. Now, towards the end of the twentieth century, we have an enormous capacity to alter our environment deliberately and at the same time produce endless side effects. Plants (and animals), as we have seen, evolve to cope with a changing environment, but the changes wrought by man are so rapid and dramatic compared to most natural changes, and on so much wider a scale, that very few plants can evolve to keep up with them. The International Union for the Conservation of Nature and Natural Resources (IUCN) estimates that some 25,000 species of plant are now so rare or endangered that they need special protection, and many of them are in imminent danger of extinction. Species of plants have, of course, always become extinct through natural failure and disaster, but the scale on which extinctions are now occurring is quite unprecedented in the history of plants, and this is brought about almost entirely by the effects of habitat destruction and alteration caused by man. For the majority of plants that become extinct, we have no knowledge at all of their nutritional, medicinal or other properties, yet extinction means that their possibilities are lost to us for ever.

At the same time, whole ecosystems are faced with extinction. Perhaps the most serious problem at present is the rapidly declining area of tropical rain forest. The figures are horrifying: nearly half of the world's tropical rain forests have already been destroyed, and the rate of destruction is accelerating. An area the size of a country three times as large as Switzerland is disappearing *every year* with virtually no replacement, and it is practically certain that all the forest will have gone by the end of the century (in little over 15 years' time) unless something radical is done. Not only is the tropical rain forest one of the most beautiful and complex of systems, but it is also one of the world's primary sources of oxygen and users of carbon dioxide. It is by no means clear what will happen when it has gone. In addition, the majority of the species found in the less accessible parts of the forest are, quite simply, unknown, and species become extinct daily before they have ever been looked at.

Nearer to home, in Northern Europe and Britain, the agricultural advances of the last few decades and the highly-subsidised (and wasteful) agricultural system has provoked a major and continuing programme of bringing 'marginal' areas into cultivation. Such 'marginal' areas are the few remaining fragments of semi-natural vegetation—the woods, the marshes, the bogs and the downlands. These fragments of vegetation have developed over hundreds or often thousands of years; they contain enormous numbers of different species living in an interacting mosaic, and they are, in effect, irreplaceable. The agricultural communities that replace them are relatively sterile, with only one or two dominant species and a few 'pests' living on them, and little else. Unfortunately, the effects do not stop there: pesticides drift into any adjoining habitats that are left, where they have insidious effects; fertilisers and pesticides drain into water courses where they kill off the less tolerant species and promote the few aggressive or tolerant ones that can stand them; and, perhaps worst of all, it means that each remaining bit of semi-natural habitat becomes more and more isolated from any comparable habitat.

So if a species becomes extinct in a wood because of a fire, spray drift, bad management, or any one of many possible reasons, it is highly unlikely that it will return. In the past, many species habitually moved around to the area that suited them best, but now the intervening distances are so great, and the land use so hostile to wild species, that this can no longer occur. The net result is that each little patch of habitat, whether it is a bog or a woodland, gradually loses its species and the whole environment becomes poorer. The basic problem lies, of course, in the fact that there are too many people. There are over 5,000 million people in the world, and the environment simply cannot sustain such an amount. In those few countries where the population is still low, such as Bhutan in the Himalayas, the natural environment is still rich and diverse.

There are a few hopeful signs, as many people and governments are at least realising the scale of the problem, but its complexity and the extent of the competing interests make any real solution unlikely.

GLOSSARY

Achene: a small non-splitting single-seeded fruit.

Actinomorphic: referring to flower structure, regular; so that it can be cut in several planes and still make two similar halves (cf. zygomorphic).

Androecium: male part of flower. Sum total of all the stamens in a flower.

Annual: a plant completing its life-cycle within one year of germination.

Anther: head of stamen, containing the pollen.

Anthocyanin: red to blue colouring dissolved in cell-sap.

Apetalous: without petals.

Apomixis: production of viable seeds without fertilisation.

Autecology: study of the life and relationships of a single plant or species.

Axil: refers to leaf; the point where a leaf joins with the stem bearing it.

Berry: fleshy fruit with seeds embedded in 'pulp'.

Biennial: plant with life-cycle normally extending over two years.

Bract: a modified leaf bearing a flower or group of flowers.

Calcareous: rich in lime, e.g. soils or water deriving from chalk or limestone.

Calyx: the outermost whorl of floral parts. Individual parts known as sepals.

Carpel: the single female 'organ' consisting of an ovary with a stigma where pollen is received.

Chlorophyll: the green pigment of plants, which is one of the main constituents in the process of photosynthesis.

Chromatid: the unseparated 'daughter' chromosome produced by chromosome division.

Chromosomes: small bodies which stain darkly, that occur in all nuclei, and which determine the inheritable characteristics of organisms.

Cleistogamy: the production of unopened flowers which fertilise themselves.

Climax vegetation: the relatively stable vegetation type that eventually occurs in any given climatic type and persists as long as those climatic conditions occur.

Corolla: the petals of a flower.

Cotyledons: the leaves within the seed—the seed-leaves.

Cryptogams: plants not reproducing by seeds, e.g. ferns.

Cuticle: a waxy layer on the outside of many parts of plants.

Dioecious: where all the flowers on one plant are either all male or all female.

Diploid: having two full sets of chromosomes (cf. haploid).

Drupe: a fleshy fruit with a hard stone.

Ecology: the study of organisms in relation to their environment.

Embryo: a young organism before separation from the parent or parental organs.

Endosperm: food reserve in the seed produced from the cell of the fertilised secondary nucleus.

Ephemeral: a very short-lived plant, with a life-cycle of much less than a year.

Epidermis: the outermost layer of cells of a leaf or young stem.

Evolution: the gradual changes that take place in organisms under pressure from environmental influences.

Fertilisation: the fusion of male and female sexual cells.

Filament: the stalk of a stamen.

Flora: the kinds of plants occurring in a given area *or* a book describing the plants of an area.

Fruit: the ripened ovary or group of ovaries; often extended to include other persistent parts of the flower such as the receptacle.

Gamete: a male or female sex organ.

Gene: the unit of inheritance.

Genetics: the study of the process of inheritance.

Germination: the early stages of growth of a seed or spore.

Glabrous: without hairs.

Gynoecium: the female parts of the flower.

Habitat: the conditions under which a plant lives naturally; often extended to mean a vegetation type.

Halophyte: a plant adapted to living in salty conditions.

Haploid: with one complete set of chromosomes (cf. diploid).

Hermaphrodite: with male and female parts, like many flowers.

Hybrid: a cross between two parents of different species.

Hypha: a thin thread of living cells, usually referring to fungi.

Hypocotyl: the stem of a seedling below the seed-leaves.

Hypogeal: refers usually to germination and means that the cotyledons remain below ground.

Inflorescence: a well-defined group of flowers, e.g. a raceme.

Insectivorous: insect-eating, i.e. able to catch insects and digest them by secretion of enzymes and absorption of the products.

Lenticel: a breathing pore of woody stems.

Meiosis: the process of cell-division in which the number of chromosomes is halved, prior to the production of sex cells.

Microclimate: climatic conditions with a very limited area, e.g. around the surface of a leaf.

Mitosis: the normal process of cell-division which replicates a cell exactly so that the offspring have the same number of chromosomes as the parent.

Monoecious: with separate male and female flowers, but both occurring on the same plant.

Multicellular: made up of several cells.

Mycorrhizal association: a specialised association between roots and fungi in the soil, e.g. in orchids.

Nucleus: (plural nuclei) a specialised part of the cell which directs many of the cell's activities.

Ovary: the part of the female organ which encloses the ovule.

Ovule: the part of the plant which may develop into a seed after fertilisation.

Parasites: plants (or animals) which live on the resources of other living organisms, e.g. mistletoe on apple trees.

Pedicel: a flower stalk.

Peduncle: an inflorescence stalk.

Perennial: a plant living for three or more years.

Petiole: a leaf stalk.

Phloem: the specialised tissue in the plant which transports foodstuffs around the plant body.

Photosynthesis: the manufacture of carbohydrates by the pigment chlorophyll from carbon dioxide and water, in the presence of sunlight.

Pollen: grains containing the male sex-cells in flowering plants. Produced in anthers.

Pollination: transfer of pollen from anthers to a suitable stigma.

Pollinium: a mass of pollen grains that stick together and may be removed *en masse* by the pollinating agent.

Polyploid: with more than 2 sets of chromosomes in each cell nucleus (cf. haploid, diploid).

Raceme: an inflorescence with stalked flowers, with the oldest flowers at the base. Similar to a spike but with flowers stalked.

Receptacle: the part of a flower stalk which bears the floral organs (i.e. the petals, carpels etc.).

Reduction division: also known as meiosis. The double division during which the number of chromosomes in the cells are halved, prior to reproduction.

Respiration: the chemical breakdown using oxygen, of carbohydrates and other organic materials in living cells to release energy and carbon dioxide.

Rhizome: a horizontal underground stem, often used for food storage.

Saprophytes: plants (or other organisms) which live on the products of dead organic matter.

Seed: a detachable structure, produced by a flowering plant which contains an embryo that can grow into an adult plant.

Self-pollination: the transfer of pollen from anther to stigma within the same plant.

Sepals: the outer floral leaves—collectively, the calyx.

Species: a group of similar organisms able to interbreed normally and produce fertile offspring. This is the basic unit of plant and animal classification.

Spike: an elongated inflorescence with unstalked flowers arising in succession on it, with the oldest at the base.

Spore: a simple reproductive unit, not differentiated into separate organs.

Stamen: the organ which produces the pollen in seed-bearing plants.

Stigma: (plural stigmata) the receptive surface of the female organs in a flower to which the pollen grains adhere before germinating.

Stoma: (plural stomata) a small hole in the leaf or young stem surface that allows the exchange of gases, including water vapour, between the inside of the organ and the atmosphere.

Succession: the series of changes which take place naturally as one vegetation type changes to another, e.g. the change from bare ground to, ultimately, oakwood.

Symbiosis: a close, mutually beneficial relationship between two organisms, e.g. between the alga and fungus that form a lichen.

Taxonomy: the study of plant classification.

Transpiration: the loss of water vapour from the surface of a plant and its uptake bv the roots to replace the lost water.

Xerophyte: a plant adapted to living under dry conditions.

Xylem: tissue consisting of dead cells hardened with lignin which transport water and mineral salts through the body of the plant. Known as 'wood' where it occurs in large amounts, as in trees.

Zygomorphic: bilaterally symmetrical, but in one plane only. Refers to flowers, e.g. the flowers of larkspur which only give 2 equal halves when cut in one particular way (cf. actinomorphic).

BIBLIOGRAPHY

Attenborough, D., *Life on Earth*, Collins/BBC, London 1979.

Beckett, G., *The Secret Life of Plants*, Octopus, London 1982.

Bellamy, David, *Bellamy on Botany*, BBC Publications 1975. A very readable introduction.

Bellamy, David, *Botanic Man*, Hamlyn, London 1978. A personal worldwide view of plants.

Brouk, B., *Plants Consumed by Man*, Academic Press, London 1975.

Clapham, A.R., Tutin, T.G. and Warburg, E.F., *Flora of the British Isles*, Cambridge University Press 1962 (& subsequent editions). The classic British flora.

Clapham, A.R., Tutin, T.G. Warburg, E.F., *Excursion Flora of the British Isles*, Cambridge University Press 1981, 3rd edition. A handy pocket edition of the above.

Davis, P. & J., and Huxley, A., *Wild Orchids of Britain and Europe*, Chatto & Windus: The Hogarth Press 1983. A well-illustrated and comprehensive guide.

Dobson, F., *Lichens, an illustrated guide* (2nd edition), The Richmond Publishing Co., Richmond 1981. One of the few illustrated guides to lichens.

Everard, B. and Morley, B., *Wild Flowers of the World*, Ebury Press, Michael Joseph 1970.

Fitter, R.S.R., *Finding Wild Flowers*, Collins, London 1971.

Harrison, S.G. et al, *The Oxford Book of Food Plants*, Oxford University Press, Oxford 1969.

Heywood, V.H. (Ed.), *Popular Encyclopaedia of Plants*, Cambridge University Press, Cambridge 1982.

Hyde, H.A., Wade, A.E., Harrison, S.G., *Welsh Ferns*, The National Museum of Wales, Cardiff 1969, 5th edition.

Martin, W. Keble, *The Concise British Flora in colour*, Ebury Press, Michael Joseph, London 1965 (and later editions). The result of 60 years' work. Still the best readily available colour illustrations of British flora.

Mitchell, A., *A Field Guide to the Trees of Britain and North Europe*, Collins, London 1974. An invaluable and comprehensive work by an expert on trees.

Owen, D.F., *What is Ecology?*, Oxford University Press, London 1974. A lucidly written introduction to ecology in general.

Pennington, W., *The History of British Vegetation*, The English Universities Press, London 1974, 2nd edition. Contains an enormous amount of research data in a readable way. Particularly strong on the study of buried pollen as a means of assessing past vegetation.

Perring, F.M. & Walters, S.M., *Atlas of the British Flora*, BSBI/Nelson, London 1962. A fascinating classic work showing the distribution of almost all British plants, with overlays on climate, geology etc. for comparison.

Perring, F.M. & Sell, P.D., *Critical Supplement to the Atlas of the British Flora*, BSBI/Nelson, London 1968.

Perring, F.M. and Farrell, L., *British Red Data books: 1 Vascular Plants*, SPNC, Lincoln 1977. A catalogue and description of the 360 or so rarest and endangered British plants. Useful, though depressing, reading.

Peterken, G.F., *Woodland conservation and management*, Chapman and Hall, London 1981. A comprehensive account of the variation in natural woodland and the ways in which it has been or should be managed.

Phillips, R., *Mushrooms and other Fungi of Great Britain & Europe*, Pan Books, London 1981. 'The most comprehensively illustrated book on fungi this century.' An excellent book with clear photographs of all fungi you are likely to find, and excellent value.

Praeger, R.L., *The Botanist in Ireland*, Hodges & Figgis, Dublin 1934. A classic work on Ireland, recently reprinted.

Ratcliffe, D.A. (Ed.), *A Nature Conservation Review*, Cambridge University Press, Cambridge 1977, Vols 1 & 2. A comprehensive study of British wildlife conservation and the best wildlife sites. A fine work, though already outdated in some respects.

Rose, F., *The Wildflower Key*, Warne, London 1981. An excellent field guide by one of Europe's best botanists. Includes keys to plants not in flower.

Tutin, T.G. & Heywood, V.H. (Eds.), *Flora europaea*, Cambridge University Press, Cambridge 1964–1980, Vols 1–5. A remarkable attempt to describe the whole of European flora.

Picture Credits

Bob Gibbons: pp. 15, 19, 23, 30, 39, 43, 50–1, 54, 63, 71, 75, 83, 86, 87, 95, 102, 107, 123, 130–1, 139
PW/Bob Gibbons Photography: pp. 78, 79, 126
RF/Bob Gibbons Photography: pp. 130 (inset), 135
Halcyon Midland Photographic Library: frontispiece, p. 6.

INDEX

Figures in italics refer to page numbers of illustrations.